# AIR
# CRASHES

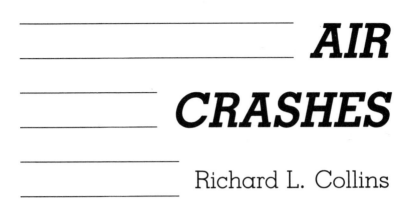

# AIR

# CRASHES

Richard L. Collins

*An Eleanor Friede Book*

MACMILLAN PUBLISHING COMPANY

NEW YORK

COLLIER MACMILLAN PUBLISHERS

LONDON

Macmillan Publishing Company
866 Third Avenue, New York, N.Y. 10022
Collier Macmillan Canada, Inc.

Library of Congress Cataloging-in-Publication Data
Collins, Richard L., 1933–
Air crashes.
"An Eleanor Friede book."
Includes index.
1. Aeronautics—Accidents.   I. Title.
TL553.5.C53   1986      629.132'52'0289      86-12808
ISBN 0-02-527150-4

10 9 8 7 6 5 4 3 2 1

Designed by Jack Meserole

Printed in the United States of America

# Contents

# Preface

Riding in or flying an airplane in ideal conditions is a nice stable way to get somewhere, enjoy sightseeing, or just to enjoy the sensations of flight. But airplanes move quickly and there's an additional dimension. When they don't wind up on a runway, touching down in a proper spot at a proper speed and in the correct configuration, the resulting transition from air to ground can be a disaster.

As a pilot, I often marvel at how well it works. Whether in Concorde, flying at a groundspeed of 1,150 knots, a few hundred miles from New York, at almost 60,000 feet, or in a light airplane at 160 knots, 50 miles from an airport at 10,000 feet, it's no mean challenge to make it all come out even on the designated runway ahead. It takes a measure of skill and judgment on the part of the pilot, to say nothing of the correct decisions that must be made along the way.

But sometimes it doesn't work. In airline flying, many of the bad accidents in recent times in the U.S. have been related to wind shear. Commuter, or regional, airlines have had some mechanical and weather problems. In general aviation—all flying excluding airline and military—the accidents are spread throughout the various phases of the activity.

When there is an aircraft accident, there's always much instant speculation on the cause and many theories surface as the newspeople talk to "experts" who may or may not know what they are talking about. In some cases, the speculation centers on something that might have gone wrong with the airplane but in most the attention centers on people. The pilot did or didn't do this or that. Or maybe it was an air traffic controller. In fact, the pilot is the key in most aircraft accidents. Even highly trained, experienced, and proficient pilots make mistakes that turn flights into disasters. This happens so seldom in airline flying that flying on the big jets of the major airlines has become one of the safest forms of transportation ever; no other form of flying is quite as good.

The reason there is an occasional slip is that it is not possible to anticipate every single thing that might happen. Even the most avid student of aviation safety would be making a pure guess if he were to project the cause of the next major disaster. The past can be examined and history used to explore probabilities but it just might not happen that way. Whatever happens, it's a cinch that the pilot did his best to get the airplane safely to the runway but wound up short.

In smaller airplanes, the pilots are often less experienced and have less training. As a result, the safety record is not as good as in the heavies. When the accident history of smaller airplanes is examined, one point stands out. Only a few of the accidents would have been prevented had the pilot been flying a heavy airplane; most would have been prevented had the pilot been trained to the same high standard and followed the same basic and methodical procedures that are used in large airplanes.

Understanding why air crashes occur is important to pilots. By following and understanding the underlying factors, we can avoid repeating the mistakes that other pilots have made. The study of accidents is, in fact, a very important part of flying safely, even though it is not incorporated in the training program pre-scribed by the Federal Aviation Administration.

When studying the subject, pilots tend to be hard on other pilots. Those who haven't crashed are, in essence, sitting in judgment of those who have crashed. And if the airplane was functioning properly up to the point that it collided with the ground, then what we seek is an idea of what led the pilot to either fly the airplane incorrectly, or to position it or handle it in such a manner as to make inevitable a crash.

Aviation has been around for a long time and is a mature business. The government agencies involved, primarily the Federal Aviation Administration and the National Transportation Safety Board, are stable, and while they are not always efficient they are conscientious and dedicated to making aviation as safe as possible. As a result, the inevitable search for a "goat," or some hint of malfeasance or wrongdoing or negligence in the wake of an accident seldom bears fruit. Usually, it's one slight error that leads to a larger error that finally snowballs into a full-fledged disaster. In this book we'll examine factors and accidents in general and in specific areas. All accidents make the papers when they occur but it's seldom that the findings of the lengthy investigations are exposed to the public. The purpose of this book, for the pilot and nonpilot, is to explore, to probe, to understand why airplanes crash.

—RICHARD L. COLLINS

# AIR
# CRASHES

# How Safe?

REALLY, NOW, how safe is flying? In years when there are a lot of major accidents, the answer appears to be "maybe not very safe." In other years, or strings of years, when there are few accidents, riding on major airlines appears tame indeed. But you always read about general aviation airplanes crashing. They don't seem to have good years.

You have to consider long-term averages when trying to assign levels of relative safety to various types of flying. The nature of the activity is such that there will be peaks and valleys in accident numbers, especially in airline flying. This activity is so highly regulated even in this age of deregulation, so much of it is done, and the airplanes used are so good and the crews so well trained, that every accident is somewhat of a fluke. And one more accident in a given year can make a big difference in the statistics for that year.

No accidents *should* happen but occasional ones are inevitable, part of the price of moving around. That is a bad thing to have to say because the goal should always be absolute safety. But this is not attainable in any form of transportation. Somewhere, somehow, at a dark and stormy time, all the bad factors are going to come together and result in an accident. The fact that they gather so seldom in some areas of flying—the operation of jet airplanes by airlines or corporations is by far the best area of risk management—proves that care and thought make flying as safe as possible.

## The Measure

How do we best measure safety, or risk, in flying? The most meaningful number to the individual is the fatal accident rate per vehicle mile. Where passenger miles are often bandied about as a measure of safety, these have little meaning to the individual. It matters not to one passenger how many other passengers are on an airplane if it crashes and all aboard perish. What we are considering, in examining the relative risk of various means of transportation, is the chance of being in a car or airplane that is involved in an accident that produces fatal injuries. The question is "What is the exposure to the ultimate risk in going from here to there?"

All the relative numbers have qualifications that will be explored. In each case, the exploration is not to make excuses for one form of transport or another. There are simply a lot of factors that go into safety records and they should be understood. Just pulling numbers out of a hat, out of context, can create a false image.

## Large Airplanes

The major scheduled airlines have an excellent safety record. Their numbers are always tiny; in an average year they might have one fatal accident for every billion miles flown. Some of the fatal accidents that do occur might not involve peril to all on board, either. There have been rare instances of, for example, an engine failing and flying debris from the engine hitting one passenger in the airplane. That's counted as a fatal accident even though all or most on board were not affected. But when one actually does crash, that is, when it comes to something other than a gentle and planned stop on a runway, the hazard is high for all people on board.

The times and places where the risk is greatest in large airline flying will be explored in separate chapters. And remember that the random risk associated with flying on major airlines is not something that an individual can do much about. Some try by avoiding airlines that they think are less safe than others. Some won't ride the national airlines of some countries because they don't feel they are up to par because they have had what appears to be an abnormal number of accidents, or because of statistics that show U.S. air carriers to have a better record than those in most of the rest of the world. Others avoid the newer "no frills" airlines because they feel that by buying used equipment and paying less to employees, they just naturally have a lower level of safety. Picking the airline you think is best can be viewed as a form of risk management but it is random. A person in the insurance industry explained how he had been having trouble placing insurance on a foreign airline's fleet but that he never had trouble with the large domestic U.S. carriers. Then he went on to say that in 1985 the people who insured Ethiopian Airlines did fine where those insuring Delta took a big hit in the accident at Dallas/Fort Worth. It *is* that random.

## Cars

How does travel on major airlines compare with automobile travel, as a point of reference? The car numbers are elusive in the sense that no statistic is available on the fatal accident rate for sober drivers wearing seatbelts and obeying all the traffic laws. We do have those options when driving. Where risk management in buying airline tickets is random and perhaps of little effect, risk management is a big deal in automobiles. Just in using the available numbers, arbitrarily reducing them to take out the huge additional risk of driving drunk, taking out pedestrians, and reducing them some more for belts and adherence to the law, you probably come up with a number that would show driving to Florida involves four or five times as much risk as flying to Florida on a major scheduled airline. It might be a little more than that, it might be a little less, depending on other factors such as the size car you drive and your basic driving skills. And where there are uncontrollable risks in driving, mainly centered on the "other guy," a sharp driver can even modify these to a certain extent by driving defensively.

## Commuter Airlines

The commuter or regional airlines, most of which fly airplanes with from 19 to 35 or so seats, don't do as well as large airlines. In fact, their fatal accident rate per vehicle mile is several times higher than the larger airlines and is probably close to the "safe driver" rate. They really should be compared with automobiles, too, because in most cases the auto is the alternative to riding on the smaller airlines. There is also a division in the regionals that should be considered. If they fly large airplanes, those with more than 30 seats, the regionals have to operate under the same regulations as the major airlines. And if the commuter airplane weighs over 12,500 pounds, the crews have to meet more stringent requirements and the airplanes have to have better performance capability after the failure of one engine. Risk management in riding on commuter airlines might thus be linked to ascertaining what kind of airplane the line uses. There is a big difference between required pilot training to fly a twin-engine piston-powered airplane with nine seats and one of the new generation turboprop airliners such as the Saab 340, ATR-42, de Havilland Dash 7 or 8, or other new ones. There's also a big difference in the capability of the aircraft. The regional airlines are making great strides in equipment, training, and risk management but there's no way they can ever consistently be as good as the major airlines because of more frequent takeoffs and landings, phases of flight where the risk is higher. But they are likely to improve and, given the effect that one accident more or less has on the record, we might well see a year in which regional airlines compile a better safety record than major airlines.

The record of the regionals can also be related to professionalism. A passenger related a tale of sloppy operation on one of the regionals bearing a big airline's name. The crew for the 19-seater showed up at the airport 15 minutes late and then went in search of the fuel truck. The pilots lounged in the terminal while the airplane was fuelled and then boarded the airplane, with the passengers, without so much as a walk around the airplane to see if anything was loose or missing. That used to be the norm in small airline operations. Now it is the exception—but it still happens and something like this reflects lack of profes-

sionalism and a generally unhealthy approach to the fine art of
flying safely. If it happens to you it might be a good idea to ask
the pilots if they shouldn't do a walk-around inspection of the
airplane.

## On-Demand Air Taxis

The airplane for charter at the local airport might be a single-
engine, a piston twin, a turboprop, or a small jet. There is a lot
lumped into this category, so any comparative rate has to be taken
with a grain of salt. But some additional risk is clear. The raw
number on the fatal accident rate per vehicle mile is probably
five or six times higher than regional airlines. For the average
user, this rate has to be reduced by a factor to take out the
accidents in cargo and small package flying, which is mostly done
at night, in all weather, with strong pressure on the pilots to make
the schedules. A person using on-demand air taxis can also do
a bit of risk management in stressing that a safe flight comes
ahead of getting there on time.

## Other General Aviation

All the rest of the flying tends to be lumped together under the
heading of general aviation. This encompasses everything from
ultralights to instructional flying, from those of us who fly our
own airplanes for transportation to corporate flight departments
operating large jets. The total accident rate per vehicle mile is

high, an estimated five or 10 times higher than regional airlines, and a great deal higher than airline flying. But the basic rate doesn't mean a lot because of the diversity of things you might choose to do within this area. There's also even more potential for individual risk management here than in driving a car. You can, for a fact, examine the frequent accident causes and avoid them, taking away much of the risk.

Starting at the top, with corporate and executive flying, where a hired crew operates the airplane to carry employees of the company that owns the airplane, the risk is relatively low, falling about on a par with commuter airlines. But there are, even within this area, some large variations. For example, when pure jets are compared with turboprops, on a per-vehicle-mile basis, turboprops had 4.6 times the risk of flying in a jet if one year's accident statistics are used as a measure of risk. Flying in the corporate jet might get the risk down close to that of the large scheduled airlines, while flying in a turboprop appears to involve more risk. There's logic in this. Corporate jet airplanes and crews meet the same requirements as the airline jet operators, and the only additional risk that could be assigned to the company jet would relate to the use of airports that are not familiar to the crew, or airports with fewer landing aids than the major terminals.

This does not brand jets as good and turboprops as bad. The higher risk in the latter is not, in most cases, directly related to the airplane. It is more directly related to the type of flying done in the turboprop and the far more relaxed regulation of the operation of that type airplane. This will be examined in detail later, but consider a few things here: to operate a turboprop (that weighs less than 12,500 pounds) requires no formal training in the specific airplane, whereas to operate a jet you must be trained, get a specific rating for the airplane, and follow a course of

recurrent training. There is no requirement to factor in performance margins on small turboprops, while this is required for
every flight in a jet. Most all jets are flown by professional crews,
and two pilots are required in most. That's not true in small
turboprops. Consider the case of the affluent pilot who climbed
to altitude in his turboprop, set the autopilot, and went to sleep.
If a couple of Air Force fighters hadn't intercepted the airplane
and awakened the pilot, the airplane would have flown to sea,
run out of fuel, and added to the higher turboprop accident rate
even though the airplane was innocent.

The record in airplanes with piston engines is not as good as
the turboprop record and is far from as good as the jet record.
But again, a lot of the risks can be managed out of any of these
airplanes. Simply meeting the crew or pilot training requirements
and performance requirement considerations of the aircraft with
the best record—the jet—can make a turboprop, a piston-
powered twin, or a simple single-engine airplane measurably
safer.

An area where the safety record is probably better than most
people think is instructional flying. Because this flying is local
in nature, more by the hour than by the mile, the accident rate
per hour is of primary importance. And it is, in fact, about as
good as the hourly rate in corporate and executive flying. How
can this be? Student pilots flying as well as the professionals in
heavier equipment? It's simply a matter of control. Risk management in instructional flying is a function of the school, the
rules, and the flight instructor managing risks both when the
instructor is in the airplane teaching the student, and when
the student is out flying solo.

## *The Rest*

Cut that student loose, though, with a private pilot's license to fly the wide variety of ultralights, singles, and twins, and the record goes downhill. It's not that light airplanes are inherently dangerous or mechanically unreliable—few of the accidents directly relate to something on the airplane failing—it's just that the system of training pilots and requiring them to maintain proficiency is sorely lacking. The accident causes are well-known and they repeat on a very predictable basis. Weather-related and stall-spin accidents lead the list and by nature the small airplane is hardly more vulnerable to these things than the large airplane. The pilots who fly the smaller airplanes are a lot more vulnerable, though, because they aren't as well trained. There is probably no endeavor where individual risk management can do more to improve the safety potential and, conversely, no area where the failure to manage risks can be more instantly lethal. The small airplane is a fine travel machine and some are excellent for recreation. It's too bad that the accident rate in this is so high but it's good that the risks can be examined as well as managed.

The question "Is it safe?" has a lot of answers. No form of transportation is totally risk free. And in aviation there are more opportunities to manage risks than in many other areas. Most fall directly in the lap of the pilot in command of the airplane—the most important element in the business of flying safely.

In examining accidents in the following chapters, and the relationships that develop between scenarios, airplanes, and pilots, the information is based on the National Transportation Safety Board's report of the accident or material released by the FAA, and is intended to bring the issues raised to the attention of the reader. It is not intended to judge or to reach any definitive conclusion about the ability or capacity of any person, living or dead, or any aircraft or accessory.

# The Pilot

DESPITE the stereotype of the pilot—slightly graying, straight teeth, and a crooked smile—they come in all sizes, shapes, and ages. Male or female, from 16 (the minimum age for solo flight) to senior citizens, there's no accurate definition of the pilot as a person. There's an equally wide range in training, proficiency, ability, and judgment. To make up for the variety, there are rules. But rules help only to outline a framework for safe operation. The real measure of success in flying comes from ability and, perish the thought of using the word, discipline. It helps to be a suspicious and curious person, too, as well as a good planner. Judgment is often mentioned as an attribute, and there's always experience to consider. Let's look at these things individually, to better understand the pilot.

## Ability

Whether it's correct or not, we all tend to judge a pilot's ability by how smoothly the airplane flies. The landing is a primary grading point. And it *is* essential that a pilot have "good hands" as they call them in the trade. The eye-brain-hand coordination that is required to fly well comes in different measures to different

11

people but it is important to understand that a "smooth" pilot is not necessarily the safest possible pilot. Indeed there is some indication that a person who finds the actual handling of an airplane quite easy might lapse into complacency. And that complacency is probably the deadliest of the flying sins. You may have read at one time or another that in many airline accidents, where there is a recorder in the cockpit to give accident investigators access to the last 30 minutes of cockpit conversation, one or more crewmembers were whistling shortly before the accident. It's been speculated that some may whistle when nervous, but it is probably more a sign of complacency. We just don't whistle and think very well at the same time—try it—so maybe whistling is a sign that the pilot feels that he has everything wired.

A smooth pilot might fly an airplane into a bad situation because he feels able to handle virtually anything that might come along. And he might fly into doubtful situations many times before reaching one that he can't handle.

Consider the regional airline pilots mentioned who didn't inspect their airplane before takeoff. Complacent, they were sure nothing was loose or missing. They might have been smooth pilots, but that might not have been good enough to save the day if something was wrong with their airplane.

None of this is to suggest that you have to be hamhanded to be a safe pilot. Far from it. An instrument approach in bad weather, for example, requires strong ability. The airplane must be made to track the instrument landing system. The airspeed must be kept on the proper value as the airplane descends through changing wind. You've heard a lot about wind shear lately, and its effect on airplanes that are landing or taking off. The accidents at Dallas/Fort Worth and New Orleans were both related to wind shear. But there's a bit of wind shear on virtually every bad-weather approach because the wind at the surface is seldom the same as the wind a thousand feet above the ground when the ceiling is low. And it takes ability to fly an airplane through even a minor wind shear while maintaining the correct airspeed and track.

Flying ability, though, still has to be considered a mechanical function. An autopilot can actually fly an airplane more smoothly and precisely than the best human pilot, and an autopilot can land and control the landing roll. What is called an autothrottle system, installed on many larger jets, can do a very precise job of setting engine power to help the autopilot maintain the desired airspeed and altitude or rate of climb or descent. But an autopilot can't think. Ability is the easiest part of the pilot's psyche to test. Ability comes from training, where a pilot learns how to manipulate the controls to make the airplane do what is desired. The testing is basically a process of seeing to it that the pilot can maintain direction of flight, airspeed, and altitude, and can do certain maneuvers within prescribed limits. The ability to conduct proper emergency procedures can be tested, as can the pilot's general knowledge of the airplane and the regulations.

Ability, or a lack of it, is a factor in a lot of general aviation accidents and far less a factor in military and airline operations where the training systems are more stringent and less likely to leave a stone unturned. Also, in general aviation flying, the requirements that a pilot maintain the ability that was demonstrated on a check ride are pretty lax. This shows up in the accident reports. For example, successfully flying a piston-engine twin after the failure of one engine is a demanding business and must be mastered when training for a multiengine rating. But there is no requirement that a pilot do anything to maintain that ability. There are a lot of crashes in piston twins after an engine fails, and the pilots flying usually have a lot of flying experience. But the ability to handle the airplane on one engine fades after training because of a lack of polishing. It's almost as if the pilot developed the ability to handle multiengine airplanes in an emergency, was then allowed to fly them, and because engines are reliable everything went well. But after a long time, one engine finally decided to fail and the pilot found that the ability to handle it was gone with time.

## Discipline

The greatest discipline in flying is found in airline and military operations. There the procedures and allowed activities are as cut-and-dried as they can get in something as diverse as flying. There is a dispatch function, where a person on the ground goes carefully over all the factors that will affect the flight and helps prepare the crew for what will be encountered during the flight. Airline and military pilots are carefully trained to use checklists and follow procedures. They aren't automated, the pilots are still human, but a great effort is put into defining the safest way to fly high-performance airplanes and the pilots are required to follow those procedures. As we'll see in some later examples, it is when this discipline breaks down that accidents occur.

The general aviation pilot flying a jet airplane is subjected to most of the disciplines of airline and military flying. A pilot has to be specially trained to fly each make and model of jet and has to take recurrent training, just as airline and military pilots must do. But pilots flying most small propeller airplanes (those weighing less than 12,500 pounds) don't have to do any of this. It becomes a rather loose arrangement where a person can be as professional about flying as he wishes. Many pilots go far beyond the legal minimums and maintain a high state of proficiency; many do not.

An example of how this works comes again with twin-engine propeller airplanes. Anyone would look at a single-engine and a twin-engine airplane and deduce that the twin should be "safer" because it has two engines. That is not the case. For a lot of reasons, the twin has proven to be more lethal after one engine fails than the single-engine airplane is after *the* engine fails. Why? Because the twin must be flown precisely—ability, as already mentioned—but the twin must also be kept away from situations where the failure of one engine would place the pilot in an impossible situation—discipline.

The "impossible situation" might come about because of the weight of the airplane, because of altitude and/or temperature, because of the available runway length, or any combination thereof. Risk management, for example, would dictate flying the airplane below maximum weight at an airport with a high elevation or on a hot day, or from a relatively short runway. Certainly it is never good practice to load an aircraft to a weight in excess of its maximum allowable takeoff weight. These are simple and cut-and-dried disciplines of airline and military flying but not of general aviation flying. A lot of airplanes and people are lost as a result, and the pilot is the key.

Single-engine airplanes don't escape the ravages of poor discipline. In a recent two-week period I was told of two separate accidents in which experienced pilots apparently attempted takeoffs from small airports in overloaded single-engine airplanes and crashed with disastrous results. What led the pilot to make the decision to attempt the takeoff? Or, put another way, why did the discipline break down?

It's tough, as an airline pilot or a general aviation pilot, to make decisions that are an inconvenience to yourself or to someone else. But that's one of the disciplines of flying. "Hey Joe, I know you want to get to East Overshoe along with that bear you just shot, but this airplane just won't lift the load. Either you go or the bear goes, but not both." That's how it has to work.

Consider the disciplines of a takeoff in a jet airplane. First it is determined that the runway length is sufficient for the airplane to accelerate to a speed called V1; as it reaches that speed it can either stop on the remaining runway or fly away safely should an engine fail. This speed is ingrained in the mind of a pilot as a place to shift from thinking of the possibility of aborting the takeoff to flying away even if one engine fails. While there is nothing in the airplane to show that the acceleration is normal and that performance will be as indicated by the book, takeoffs tend to be very precise affairs. Given that everything works well, each takeoff should be as per the book. Additionally, safety margins are built into the performance data to give a cushion.

Next place yourself in the position of the captain of the ship during a takeoff roll. Everything is going well—or is it? What would you do if the first officer said, "That doesn't seem right, does it?" If you were the captain, would you abort the takeoff? There's really not a lot of time to think about it; it's something that a pilot would have to do as a discipline. If anyone or anything suggests the slightest doubt, stop and check it out. I was riding in a 747 one day when the crew did just that. We were thundering down the runway; suddenly the power was chopped and the big airplane was brought to a stop. After some ground checks (but no maintenance work) we took off. The captain just wasn't sure on that first go, so he stopped.

In another example, a friend who was flying 747s experienced a spectacular failure of one engine right at V1, the decision speed, on a maximum-weight takeoff. He was accompanied by a check captain in the right seat because it was his first run over this route. Two captains, and the failure occurred just before the captain in the right seat called V1. The takeoff was aborted and the aircraft was barely stopped within the confines of the airport. Braking was so heavy that a number of tires blew out after the stop, due to overheating.

While my friend said that they might have done better to go ahead and take off, and circle and come back for a landing (after dumping a lot of fuel to lighten the load), the split-second nature of the event dictated that the procedure—abort in case of failure before V1—had to be followed. That takes discipline. Bang, airspeed below V1, stop. Bang, airspeed above V1, go.

### Suspicion and Curiosity

In developing the ability to safely operate an airplane, suspicion and curiosity are real virtues. Morbid as it sounds, the pilot who feels that the airplanes, airports, and skies of the world are full of lurking hazards, things intent on killing anyone who isn't on top of everything, is likely the safest pilot.

How might this work? Relate it, for example, to a thunder-shower near the destination airport. A pilot lacking curiosity might look at the shower, dismiss it as small, and press on. A suspicious and curious pilot would go through the whole exercise of determining that this storm is not the one with his name on it. While no two thunderstorms are exactly alike, the general mechanics of thunderstorms are well known and a curious pilot will look at all with great suspicion and will work out the worst-case scenario.

To get a little more technical, I'll relate a tale about suspicion and how it can be beneficial to safe flying. I was in my airplane one stormy day, flying from Wichita, Kansas, to Indianapolis,

Indiana. My son, an experienced pilot, was flying the airplane and we were approaching an area of rain and thunderstorms that was about 200 miles wide. There were sigmets (advisories of significant meterological phenomenon) for thunderstorms in the area. The airplane had an airborne weather radar that displays rain, as well as a Stormscope, a device that plots and displays electrical discharges in the atmosphere—lightning.

If I had been a little more suspicious in advance this little drill would have worked better. We were flying at Flight Level 190 (19,000 feet) where the temperature was below freezing. On entering the clouds, then, the airplane would collect ice. It did, including ice on the radar antenna housing. This compromises the signal and so I had to become doubly suspicious of any weather return displayed on the screen. Rain would show as lighter in intensity because of the ice on the radome. The lightest rainfall rate displayed is shown in green, and I think my son thought I was being overly cautious when I told him to turn left and go around an area of green return that was ahead. But the suspicion was that it might not be green, or light rain, if the radome didn't have ice on it.

It pays to be suspicious of information, too. On my airplane, for example, the performance specifications show that the airplane will take off on a hot day and climb to an altitude of 50 feet off a runway that is 2,500 feet long. No way I'd try that because I'm suspicious of the numbers and would add a margin.

How to combine suspicion with discipline for safer flying? This has to come from making a decision to solve anything that is suspicious as soon as possible. Airplanes move fast. Quickly doing something about suspicions is required of pilots.

To examine this, consider the cockpit conversation between

the captain and the first officer of the Air Florida 737 that was lost on takeoff at Washington National Airport. The conversation starts as the airplane begins its takeoff roll. The first officer was flying (from the right seat).

CAPTAIN: Holler if you need the wipers.
CAPTAIN: It's spooled.
UNIDENTIFIED: Really cold here.
FIRST OFFICER: Got em?
CAPTAIN: Real cold.
FIRST OFFICER: God, look at that thing.
FIRST OFFICER: That don't seem right, does it?
FIRST OFFICER: Ah, that's not right.
CAPTAIN: Yes it is, there's eighty.
FIRST OFFICER: Naw, I don't think that's right.
FIRST OFFICER: Ah, maybe it is.
CAPTAIN: Hundred and twenty.
FIRST OFFICER: I don't know.

They continued the takeoff, were unable to keep the airplane aloft, and subsequently hit the 14th Street Bridge and then the Potomac River. The NTSB's probable cause was ". . . the flight-crew's failure to use engine anti-ice during ground operation and takeoff, their decision to take off with snow/ice on the airfoil surfaces of the aircraft, and the captain's failure to reject the takeoff during the early stages when his attention was called to anomalous engine instrument readings." In the conversation related, the first officer was the one with doubt. The captain would have to make the decision. But there's not a lot of time to think. From the start of the conversation until they were past a point where the takeoff could be aborted took less than 30 seconds. The only way for a pilot to save the day in such a situation is with the discipline to stop if anything looks suspicious.

## Wrong Runway

A lack of curiosity or suspicion led to a bizarre accident at Anchorage, Alaska, one foggy winter day.

A Korean Air Lines DC-10 was departing from Anchorage on a cargo flight to Los Angeles. At the same time, a SouthCentral Air Piper Navajo was departing on a scheduled commuter flight to Kenai, Alaska. The Navajo was cleared to Runway 6L, the DC-10 to Runway 32.

The Navajo required 1,800 feet of runway visual range for takeoff and when he reached a point just short of Runway 6L, the air traffic controller told him, "It's not quite there yet, we got a thousand, I'll let you know when it comes up."

The DC-10 was given the choice of Runway 6R or 32, where it would require a quarter of a mile visibility for takeoff. The crew elected to take 32 even though the tower was reporting visibility of ⅛ of a mile at the time. The tower could not observe either airplane. The DC-10 captain made some required position reports while taxiing and then reported ready for takeoff. The flight was cleared for takeoff on Runway 32 and shortly thereafter the Navajo was cleared onto Runway 6L, to hold awaiting takeoff clearance. The runway visual range was up to 1,800 feet by this time.

Two minutes and 18 seconds after being cleared for takeoff the KAL DC-10 crew reported that they were starting the takeoff roll. But they were on the wrong runway.

The pilot of the Navajo made the following statement: ". . . heard them clear KAL for departure. . . . 30, 40 seconds later saw headlights down the runway. . . . truck on runway? . . . lights got bigger and bigger and kept going faster and faster. . . . ducked below cockpit and told passengers to do the same. . . . we felt impact."

The KAL DC-10 had taxied onto Runway 24R (which is 6L going the other way) only 2,400 feet from the end of the runway and attempted a takeoff. It hit the Navajo at the end (miraculously, everyone survived) and then traveled 1,400 feet more before stopping. The takeoff wouldn't have been successful even if the Navajo hadn't been there because the calculated required runway length for the DC-10's takeoff was 8,150 feet. After 2,400 feet, it was just getting started.

Because runway numbers reflect the magnetic bearing of the runway, and because in every cockpit there is plenty of information on the magnetic heading of the airplane, how could a flightcrew attempt a takeoff on Runway 32 with all the heading indicators on 24?

The captain of the KAL DC-10 stated, in part: "I left the north ramp at 1357. I was instructed to taxi to Runway 32, and I turned the aircraft to the left. I could not see the yellow taxiline, so I turned slightly to the right, attempting to see the taxiline. I saw the line very dimly through the heavy ice fog. While I was concentrating heavily on following the line, the tower advised me to go on to the east-west taxiway. I thought I saw the taxiway on my right and turned to the right onto it. The visibility was so poor that it was difficult to see the taxiway markings. I continued to taxi and my copilot (the first officer) confirmed that the north-south taxiway was to the right. At that time, we informed the tower that we were entering the east-west taxiway. The tower then instructed us to hold short on 32 holding point. We thought Runway 32 was to the right of the aircraft. The tower then told us to taxi into position and hold. I turned right, entered

Runway 32, and stopped. Due to the poor visibility, I felt unsure that the aircraft was on the correct runway. I looked for identifying markings, but could not see any. I discussed this with my copilot, who felt sure we were on the correct runway. After three to four minutes of discussion, I considered taking Runway 6R because of my uncertainty. However the runway size and lighting appeared to be correct, so I decided to take off. Six to seven seconds after beginning my takeoff, I saw the other aircraft directly in front of me. I knew that a head-on collision would be fatal for the people aboard both planes, so I turned slightly to the left and lifted the nose of my aircraft. A moment later, I felt and heard the crash."

The National Transportation Safety Board determined that the probable causes of this accident were ". . . failure of the pilot of Korean Air Lines Flight 084 to follow accepted procedures during taxi, which caused him to become disoriented when selecting the runway; the failure of the pilot to use the compass to confirm his position; and the decision of the pilot to take off when he was unsure the airplane was positioned on the correct runway. Contributing to the accident was the fog, which reduced visibility to a point that the pilot could not ascertain his position visually and the control tower personnel could not assist the pilot. Also contributing to the accident was a lack of legible taxiway and runway signs at several intersections passed by flight 084 while it was taxiing."

The causes of this accident were simple and the potential for a similar accident exists at many airports around the world every day. Prevention would be possible with suspicion and verification that the runway is the correct one—easily done with the compass.

The pressure on a pilot to continue even if something suspicious arises can't be discounted. After there is a long wait for a takeoff and the time comes to go, the pressure would be to go if at all possible. Aborting the takeoff would perhaps mean a trip back to the ramp for service and then another long wait for takeoff clearance. In the case of an airline pilot, this would have a ripple effect on schedules all along the line. For a pilot flying a corporate jet, it might mean the boss misses the appointment. For a businessman-pilot it might mean missing a big deal. For a pilot flying for personal reasons, it might mean missing a desired liaison or being late getting back to work. We'd like to think that none of those things would affect a pilot's decisions, whether they regard unusual noises or instrument indications on takeoff or anything else. But human nature is human nature and pilots have to learn to handle this by following procedures and the discipline of reacting to anything suspicious.

Being curious and suspicious brings something else to flying. That's total attention to the airplane while flying. A lack of curiosity about everything that is going on can let the old brain go into coast, which is bad. Or, if it doesn't go into coast, it can wander to other things. There have been cases where cockpit voice recordings have overheard pilots talking about other things and events while the airplane is headed toward a crash. If pilots are less complacent and more curious about everything that is going on in relation to the airplane, thoughts are not likely to wander. Some airlines have instituted a firm policy of no distraction of the cockpit crew by the cabin crew when below 10,000 feet (except in case of a cabin emergency), and no cockpit conversation about anything other than the operation of the aircraft when below 10,000 feet. That is a good discipline because it tends to enforce concentration on the task at hand. (In light airplanes, which often cruise below 10,000 feet, the discipline might be applied while the airplane is climbing or descending and relaxed while it is cruising en route.)

An example of how cockpit banter can precede trouble is found in the following conversation between the captain and copilot of an airliner as it approached for landing.

UNIDENTIFIED: Are we going in on ILS or something like that?

CAPTAIN: Ah, something like that, yeah it's . . .

UNIDENTIFIED: Something like that?

CAPTAIN: Yeah, it probably won't resemble anything you've ever seen on an ILS.

[Laughter]

UNIDENTIFIED: It doesn't look that bad.

CAPTAIN: All right . . . beautiful.

CAPTAIN: Did you ever notice when you need those spoilers, they're never there.

UNIDENTIFIED: The what?

(Laughter)

FIRST OFFICER: Two people doing what?

FIRST OFFICER: He's kidding.

CAPTAIN: What kind of car?

FIRST OFFICER: Heh heh . . . station wagon.

CAPTAIN: Oh, partial obscuration, in other words, we'll be able to see it until we get down to the ground.

FIRST OFFICER: Yeah, that's the story of my life. Four miles from Briggs.

CAPTAIN: Why don't we get . . . Where's my highway? That's what I'm looking for.

FIRST OFFICER: I got an Esso road map right here, captain, it's got all the airports on it.

CAPTAIN: I got a road map, but I don't have the highway I wanted. . . . Oh well.

UNIDENTIFIED: Well, it looks like any other city.

CAPTAIN: The tower is what? One oh nine something or other.

FIRST OFFICER: One zero nine point five. Six degrees inbound.
CAPTAIN: All right.
FIRST OFFICER: Marker at twenty-five hundred.
CAPTAIN: Six and twenty-five thou . . . it inbound.
CAPTAIN: I hate to see . . . that evening sun . . . [singing]
FIRST OFFICER: Why don't they fix these things?
UNIDENTIFIED: Okay.
CAPTAIN: Cleared for the approach he said?
FIRST OFFICER: PA system, funny . . . no.
CAPTAIN: He didn't say?
FIRST OFFICER: He turned, he told us to turn right.
CAPTAIN: Ohh.
UNIDENTIFIED: Nah.
FIRST OFFICER: . . . right up, huh . . .
UNIDENTIFIED: I don't know [laughter].
CAPTAIN: Let's have the gear down, please.
[Laughter]
CAPTAIN: That was funny.
FIRST OFFICER: And the markers are turned on if that is what you'd like.
CAPTAIN: All right . . . landing final checklist.
FIRST OFFICER: And the no smoking on . . . once around the beads.
CAPTAIN: What happened, how come I'm . . .
FIRST OFFICER: Ignition control override . . . radar's up and off . . . gear down, three green, annunciator panel checked . . . spoilers.
CAPTAIN: They're armed.
FIRST OFFICER: Flaps, slats, flags scanned. On the glideslope, what about that?
CAPTAIN: I hate to see the evening [singing] . . . Just like the simulator, ain't it? . . . Outer marker, all's well.
FIRST OFFICER: Sounds like you are working on something.
CAPTAIN: Good show. . . . Okay, keep a sharp eye out.
FIRST OFFICER: I will . . . sight us an aircraft, an airport or something.
CAPTAIN: You're right about the wind.

FIRST OFFICER: Yeah . . . fast.

CAPTAIN: Slow with that speed . . . that speed command is some kinda bad.

FIRST OFFICER: Stick with my system.

CAPTAIN: Yeah [*laughter*] . . . still a little high and fast, though, but that's not . . . that . . . that's not the problem.

FIRST OFFICER: Yeah.

[*Sound of female voice*]

UNIDENTIFIED: A loading problem.

FIRST OFFICER: Three hundred feet . . . above the glideslope . . . minimums . . . the airport's on your left.

UNIDENTIFIED: Fantastic.

FIRST OFFICER: We're not stopped yet.

UNIDENTIFIED: What?

CAPTAIN: We're not stopped.

FIRST OFFICER: You're not gonna stop.

CAPTAIN: Anti-skid on.

FIRST OFFICER: One hundred.

CAPTAIN: Anti-skid.

FIRST OFFICER: Anti-skid.

UNIDENTIFIED: It's going off the end.

The airplane did indeed go off the end of the runway and in its probable cause statement about this accident, the National Transportation Safety Board cited "the captain's decision to complete the landing at an excessive airspeed and at a distance too far down a wet runway to permit the safe stopping of the aircraft. Factors which contributed to the accident were: (1) Lack of airspeed awareness during the final portion of the approach, (2) the erroneous indication of the speed command indicator, and (3) hydroplaning." Consider from the transcript what might have led a professional flightcrew to make these errors. You can't say that lighthearted banter might be the cause of an accident but it is not a reflection of suspicion that the approach ahead is the most important one of a career. Each flight is actually that—the most important one—and the pilot who suspects that is more likely to park the airplane at the gate without incident.

## *Planning*

The ability to plan ahead is a vital pilot skill. This includes careful preflight planning, to make certain the proposed flight isn't a disaster looking for a place to occur, and it becomes even more critical when aloft. A pilot has to have the ability to look at the current position and condition of the airplane and make a continuous plan for everything to come out as desired. I was flying with an Air Force instructor in a T-38 one day and he told me how challenging it is to teach new, relatively inexperienced pilots how to fly precisely in supersonic airplanes. You have to be satisfied that they are thinking about the right thing as well as making sure they are thinking about what to do next and after that, and after that, and after that—all the way to the chocks on the ramp.

Not only do pilots have to plan ahead, they have to have plans for every eventuality. Going back to thunderstorms again, if a pilot opted to fly through one on approach, he'd need a plan. The airflow patterns in and around storms have general characteristics that are known, so a pilot can make a plan to fly the airplane in a way that will minimize the effects of the changing wind. That doesn't mean it can work every time, though, because thunderstorms come in all strengths and some produce wind shear that can defeat even the best plan. Planning has to consider that fact, and the correct plan might be a diversion or a delay.

## *Judgment*

Much has been made over the role of judgment in flying, but the word seems out of place to me in this context because it is usually thought of as something that a person does or does not possess on a full-time, permanent basis. It is not generally thought of as something that can be taught. A pilot needs to think in terms of things that can be learned and controlled, and of things that can and cannot be done with airplanes. To apply a mystical "judgment" measure to the situation doesn't really work. A person who seems to possess good judgment might not be a good pilot; on the other hand a person who seems downright flaky in some things might be fine in an airplane.

## *Experience*

The stereotypical "good pilot" has a lot of experience, right? Sort of right, maybe. Experience is of value only if it relates to the chore at hand. Experience can also be knowledge, but here it is of value only if used properly. Certainly a person who has seen many different weather situations is ahead of a pilot who is being rained on for the first time. But if the first-timer has studied and is up on weather, and its effect on airplanes, he can do okay. Certainly there is a broad general experience among pilots and it is possible to learn from others. To me, as valuable as experience is the study of the mistakes of others, the accident reports, the bad experiences. Some outside observers must think that pilots are a ghoulish bunch because of the way we pore over the mistakes of others, but this is really a curiosity that is very productive. The chain of events that led to the downfall of one pilot can be salted away in memory. Then if some dark and stormy night you sense that the chain is starting, perhaps it can be arrested in time through a change in course or plan of action. The requirement is for pilots to look at accident reports with the thought, "There but for the grace of God go I." It doesn't help if they are studied with the thought, "I'd never, ever do anything like that."

A lot of flying hours doesn't insulate a person against the hazards of aviation. Some people can fly sloppy for a long time, only to come a cropper the first day everything doesn't go right. On the other hand, there are many pilots who are very good with a limited amount of flying time. The military has a lot of these. I have flown in formation with military student pilots flying T-38 supersonic trainers and marvelled at how well they are able to fly with but 150 hours of flying experience. But that 150 hours is backed up with a full-time training commitment on their part. They are living and breathing aviation, full-time, all day and all night. They are also basically very bright (or they wouldn't be in military flight training) and well-coordinated. When they get wings (with 175 total flying hours) they are highly trained and well-equipped to go out and get experience. On the other hand, a civilian pilot can muddle through the training for a private pilot license, learn only what he is forced to learn to meet the minimum requirements, and still be a pretty lousy pilot at the 175-hour point. Moving up, a pilot with 15,000 hours is thought of by some as being practically invincible. He is not. There are a lot of 15,000-hour pilots in the accident report files. The key is not in how many hours a pilot has flown, the key is in the pilot understanding that the next hour or flight is far more important than all those successfully completed in the past.

## The Effect of Outside Influences

We like to think of the pilot as being a person who is totally dedicated to flying as he combines ability, discipline, suspicion, curiosity, and experience to fly safely. But even the most dedicated pilot can be affected by outside influences. In general aviation flying, businessmen-pilots and those flying for personal reasons have a lot of things to think about other than airplanes and flying. Many airline pilots have outside business interests, and military pilots flying in the National Guard or Reserve units have full-time civilian jobs.

Outside influences don't have to bring a negative effect to flying. A well-rounded person with a lot of interests should do better at everything, as long as he thinks about the proper thing at the proper time.

Some airlines try to regulate outside influences by prohibiting certain actions—usually other flying—by crewmembers. Others, mostly outside the U.S., try to discourage outside business activities. But by and large, most pilots have many other interests and outside influences, and there's nothing wrong with this as long as they are separated from flying.

I have had the interesting pleasure of flying with a wide variety of pilots—from talented but relatively inexperienced military pilots to airline crews that fly airplanes up to Concorde with total precision, from airline pilots who are a little on the sloppy side to general aviation pilots who cover the spectrum from superb airmen to real honest-to-gosh uncoordinated bumbling klutzes—as well as flying a lot myself. And while I don't subscribe to the theory that most accidents can be blamed on the pilot in the strictest sense of the word, I do subscribe to the theory that in most accidents different action on the part of the pilot could have prompted a better outcome.

To be sure, there are accidents that are a direct result of something on the airplane breaking. In a few of those there was little the pilot could have done to change the outcome. Once the failure occurred and the airplane was damaged or its systems disabled, the pilot had become a passenger. There are other accidents in which something failed and the pilot had a chance but didn't take advantage of that chance. In others, accident patterns develop in a specific phase of flight or in specific airplanes and the patterns are ignored by some pilots, leading to an unnecessary continuation of the pattern.

There is, in this day of consumer activism and incessant litigation, a growing propensity to try to reduce the role of the pilot as a cause of accidents. It's trendy to single out other things and often where the NTSB has found an accident to be the result of pilot action, the courts hold the manufacturer, the airline, or the government responsible. Without getting into the right or wrong of that, let me say this holds the potential of weakening pilot skill as well as responsibility. It almost never actually happens that poor Joe Pilot was the victim of the airplane. Instead, the airplane and passengers are more often the victim of something done or not done by poor Joe Pilot. When an airplane design is brand new, it might hold a few secrets and spring a few surprises. But the hours flown (and accidents had) in newly designed airplanes are tiny when compared with the activity in mature designs, designs that keep no secrets from the suspicious pilot. The person in the left front seat is, without doubt, the captain of the ship, and the U.S. regulations clearly define the role: "The pilot in command of an aircraft is directly responsible for, and is the final authority as to, the operation of that aircraft." That's pretty clear.

# 3

## The Rules, Regulations, and Procedures

THE RULES, REGULATIONS, AND PROCEDURES under which people fly and people or companies operate airplanes should be, and for the most part are, guidelines to relatively safe operation. I say "relatively" because everything that is legal to do in an airplane is not necessarily safe. The fact that many of the rules came after an accident means that accidents of a type that have not yet happened are not prohibited by law—and flying is complex enough that everything that can happen has not yet happened. There can still be surprises. The good pilot devotes a lot of time to avoiding surprises when flying—they have no place in airplanes—but it takes much more than knowledge of the rules to do this. The pilot has to have a broad knowledge of airplanes and the atmosphere as well as the traits outlined in chapter 2.

There are a number of examples of accidents leading to a change in the way we fly; perhaps the most classic example of all was TWA Flight 514, a Boeing 727 that was inbound to Washington, D.C. The flight was scheduled to land at Washington National Airport but, because of a storm that was pummelling the area, the destination was switched to Dulles International Airport, west of Washington. Surface winds were strong and the crosswind at National was excessive, so no large air carrier airplanes were using that airport.

There was no instrument landing system on Runway 12 at Dulles, the one most nearly aligned with the surface wind, and the crew was told to expect a VOR/DME approach to Runway 12. This would give only directional guidance, leading the airplane to the runway, and the altitude flown would be up to the crew and the air traffic controller. The airplane was flying level at an assigned altitude of 7,000 feet when the controller said, "TWA 514, you're cleared for a VOR/DME approach to Runway 12." The captain acknowledged the clearance and there followed a flight deck discussion about what altitude should next be flown.

The captain said, "You know, according to this dumb sheet it says thirty-four hundred to Round Hill is our minimum altitude." The flight engineer then asked where the captain saw that and the captain replied, "Well, here. Round Hill is eleven and a half DME." The first officer said, "Well but—" The captain replied, "When he clears you that means you can go to your—" An unidentified voice said, "Initial approach," and another unidentified voice said, "Yeah!" Then the captain said, "Initial approach altitude." The flight engineer then said, "We're out a—twenty-eight for eighteen." An unidentified voice said, "Right," and someone said, "One to go." (The dumb sheet referred to was apparently the instrument approach chart; the words attributed to an unidentified person came from one of the flightcrew members.)

The flight data recorder readout indicated that after the aircraft left 7,000 feet, the descent was continuous down to an altitude of 1,750 feet where it varied a bit because of the turbulent air. The aircraft flew into the west slope of Mount Weather, Virginia, about 25 miles from the airport, at an elevation of 1,670 feet.

High terrain was depicted on the approach chart but the crew was obviously convinced that when the controller cleared them for the approach it was okay to descend to the 1,800 feet depicted as a minimum altitude for the approach segment from Round Hill on toward the airport.

The National Transportation Safety Board report on this accident stated that the crew's decision to descend was a result of inadequacies and lack of clarity in the air traffic control procedures, which led to a misunderstanding on the part of the pilots and of the controllers regarding each other's responsibilities during operations in terminal areas under instrument meteorological conditions. The NTSB castigated the FAA for failing to take timely action to resolve the confusion and misinterpretation of air traffic terminology although the agency had been aware of the problem for several years. Indeed, at least one other aircraft had descended as did TWA 514, though it descended a little later, cleared the terrain, and landed safely.

## Shut the Barn Door

As a result of this accident, the FAA instituted a new procedure where the controller specifies an altitude to be maintained until reaching a certain point. Instead of just clearing an aircraft for a VOR/DME approach to Runway 12, they started adding something like this: "Maintain 4,000 until passing Round Hill." That was a direct response to the TWA 514 accident and every time an airplane is cleared for an approach, there's that restriction in the clearance.

The air traffic control system has a lot of its roots in accidents. The collision over the Grand Canyon between a TWA and a

United Airlines airplane on June 30, 1956, set in motion a vast revamping of the air traffic control system. Now, all aircraft flying above 18,000 feet are required to fly under instrument flight rules and are positively controlled from the ground. A series of midair collisions between light airplanes and airline jets in the '60s prompted the development of terminal control areas, TCAs, at major airports where all traffic is controlled and separated. It's more than ironic that none of the airports near which the collisions occurred—Indianapolis, Indiana; Dayton, Ohio; and Asheville, North Carolina—has a TCA. The traffic at those airports just isn't sufficient to justify one. San Diego, where there was a collision between a Pacific Southwest Airlines 727 and a Cessna 172 in 1978, did get a terminal control area.

## Tale of Two Systems

The fact that there are really two "systems" of air traffic control means that there will inevitably be some conflict. Virtually all airline aircraft, most of the higher performance general aviation aircraft, and almost all military aircraft fly by instrument flight rules all the time. That's one "system." Many of the lower performance general aviation aircraft fly by visual flight rules, where the "system" consists primarily of the pilots flying in good weather, avoiding areas of positive control airspace, and looking for and avoiding other traffic visually. "See and avoid," as it is called, is a procedure that has served reasonably well but the collision at San Diego illustrates its limitations.

The PSA 727 was arriving, and while it was on an IFR flight plan, the aircraft was making a visual approach in good weather. The crew of the 727 was advised of the position of the Cessna;

they acknowledged, reported that they had the aircraft in sight, and were told to maintain visual separation from the aircraft. Control of the flight was transferred from approach control over to the tower, and when the 727 reported on its downwind leg the tower again advised the airline crew of the Cessna's position. At this time, they had apparently lost sight of the other aircraft; the crew thought they had passed it and continued the approach. Starting when the aircraft was still talking to approach control, the transcript of the cockpit recorder tells the sad tale. Intracockpit conversation is italicized.

APPROACH CONTROL: PSA 182, traffic's at 12 o'clock, three miles, out of 1,700.

FIRST OFFICER: Got 'em, traffic in sight.

APPROACH CONTROL: Okay, sir, maintain visual separation. Contact Lindbergh Tower, 133.3. Have a nice day now.

CAPTAIN: Lindbergh PSA 182 downwind.

TOWER: PSA 183, Lindbergh Tower, ah, traffic twelve o'clock, one mile, a Cessna.

CAPTAIN: *Is that the one we're looking at?*

FIRST OFFICER: *Yeah, but I don't see him now.*

CAPTAIN: Okay, we had it there a minute ago.

TOWER: 182 Roger.

CAPTAIN: *I think he passed off to our right. . . . He was right over here a minute ago.*

FIRST OFFICER: *Yeah.*

FIRST OFFICER: *Are we clear of that Cessna?*

ENGINEER: *Suppose to be.*

CAPTAIN: *I guess.*

FOURTH CREWMEMBER: *I hope.*

CAPTAIN: *Oh yeah, before we turned downwind, I saw him about one o'clock, probably behind us now.*

FIRST OFFICER: *Gear down.*

FIRST OFFICER: *There's one underneath.*

FIRST OFFICER: *I was looking at that inbound there.*
CAPTAIN: *Whoop.*
FIRST OFFICER: *Aghhh!*
[*Sound of impact.*]
FOURTH CREWMEMBER: *Oh @#$%!*
CAPTAIN: *Easy baby, easy baby. . . . What have we got here?*
FIRST OFFICER: *It's bad.*
CAPTAIN: *Huh?*
FIRST OFFICER: *We are hit, man, we are hit.*
CAPTAIN: Tower, we're going down, this is PSA.
TOWER: Okay, we'll call the equipment for you.

The PSA 727 crashed into a residential area, killing all 135 people on board plus seven persons on the ground. The two people in the Cessna were also killed.

A study of cockpit visibility showed that the Cessna would have been centered in both the captain's and first officer's windshield from 170 seconds to 90 seconds before the collision. Thereafter the airplane would have been in the lower portion of the windshield, just below the windshield wipers. Because the 727 was overtaking the much slower Cessna, the large airplane was hidden behind the Cessna's ceiling structure for most of the time before the collision.

There are rules and procedures covering all this and the NTSB quoted many in its report. One states, "Each aircraft that is being overtaken has the right of way, and each pilot of an overtaking aircraft shall alter course to the right to pass well clear."

Such rules can be used and interpreted to say who should have done what, but once the 727 crew lost sight of the Cessna— they were viewing it against the background of a city, making it quite difficult to see—the potential for a collision existed. The report on the accident found the probable cause to be the failure of the flightcrew of Flight 182 to comply with the provisions of

a maintain-visual-separation clearance, including the requirement to inform the controller when they no longer had the other
aircraft in sight. Contributing to the accident were the air traffic
control procedures in effect that authorized the controller to use
visual separation procedures to separate two aircraft on potentially
conflicting tracks when the capability was available to provide
either lateral or vertical radar separation to either aircraft. In other
words, the NTSB questioned the use of "see-and-avoid" where
the capability exists to positively separate all traffic.

## Puzzle

The random nature with which midair collisions occur adds a
perplexing piece to the air traffic control system puzzle. For years
there were almost no collisions between high-performance
turbine-powered general aviation airplanes and light airplanes.
Then there were three in just a couple of months. The one that
got the most press was between a four-place Piper Archer and a
trijet Falcon 50 near Teterboro Airport in New Jersey. In this
one, each airplane had been told of the other by air traffic control
but they still collided. The report on the accident hadn't been
issued as this was written but it might come out remarkably similar
to the PSA-Cessna crash. Another collision during this time was
between a two-place Cessna and a turboprop Cessna, near Love
Field in Dallas. These two collisions might lead you to think
that it's in the major metropolitan areas where these things always
happen, but the third one in the series occurred at the relatively
quiet airport operated by Auburn University, in Alabama. A
landing Learjet and a departing ultralight airplane collided; the
Learjet rolled inverted and slid to a stop. The Learjet copilot and

ultralight pilot died in the accident. Collisions don't all occur where there is a lot of air traffic.

To its credit, the FAA recognized early on that the mix of fast and slow airplanes could be lethal and it set out to make rules to minimize the conflict. There is a 250-knot speed limit below 10,000 feet, where most of the VFR traffic is concentrated. That gives everyone more time to see and avoid. There is a 200-knot speed limit in the airport traffic area (within five miles of an airport with a control tower, up to an altitude of 3,000 feet above the ground), and the same speed limit is imposed on aircraft flying beneath a terminal control area. Piston-engine-powered airplanes are limited to 156 knots in the airport traffic area.

## More Radar

The FAA is also implementing airport radar service areas around busier airports served by airlines. In these areas all aircraft are required to be in contact with air traffic control.

Another program resulted from one of the '60s collisions, the one at Indianapolis. In that accident the jet was descending to quite a low altitude a substantial distance from the airport. It was basically flying in airspace where the pilot of a light airplane wouldn't logically expect a jet to be. After this collision, the FAA started a "keep 'em high" procedure to keep jets as high as possible for as long as possible to minimize the low-level conflict. But what happened to this is an example of how anybody, or any government agency, can slip back into habits that have proven to cause a problem.

"Keep 'em high" was (and is) in some areas difficult to implement because of the relationship between arriving and departing airplanes in a major terminal area. An accepted procedure is to either run the arrivals under the departures, or vice versa. The former method seems preferred at most locations so arriving airplanes are again reaching relatively low altitudes quite a distance from their destination airport. One procedure that was in use in 1985 for arrivals at Newark, New Jersey, put airline jet aircraft as low as 6,000 feet 30 miles away from the terminal control area and almost 50 miles away from the airport of intended landing in an area where light aircraft traffic is heavy. Because the FAA's original plan was to keep airline aircraft generally above 7,000 feet until they descended into the TCA, and because few light-airplane pilots would expect to find a jet flying that low so far from any airline airport, such a procedure carries with it an increased exposure to midair collision risk. How much more? Who knows. It is a random thing.

"See-and-avoid" still has viability, and where light airplanes have transponders that report both position and altitude to the controller whether or not the pilot is in radio contact, meaningful traffic advisories can be given to the airline crews. But, as shown in the PSA/Cessna collision, and as possibly was the case in the Teterboro collision between the Falcon 50 and the Piper, giving a traffic advisory doesn't necessarily preclude a collision. And where air traffic controllers are supposed to issue all the traffic that they see on the scope to airliners, at busy times they do miss occasional calls.

There is a technological answer in the wings that many think will be a big step forward in reducing the number of midair collisions. It's an electronic collision avoidance system that will give flightcrews word on other aircraft that might be a threat. This has been under development for a number of years, and while it can't be considered a cure-all, it does hold some promise.

Since the rash of collisions between airliners and small airplanes in the '60s, the things the FAA has done have to be considered in a positive light. The incidence of collisions through the '70s and into the mid-'80s has been far less and there are rules and procedures under consideration that should further reduce the risk.

## Weather

Regardless of what type flight operation is being conducted— airline, general aviation, or military—when there is an accident there is often some question about the weather. The relationship between the pilot wanting to complete the flight and the bad weather is often considered. Whether an airline pilot is trying to make a schedule, a business pilot is trying to make a big deal, or a military pilot is trying to get home, the question is always, "Did the pilot try too hard to complete the flight?" It's an interesting area, the biggest safety problem, and there are regulations and procedural answers for some questions but no answers for others.

### ANSWERS

A common scenario involves an airplane flying into the ground while making an instrument approach in bad weather. This type of accident seldom involves airlines flying jet equipment because most of the runways that they use have full instrument landing systems with glidepath guidance. Regional airlines have some accidents like this, general aviation has a lot. The basic rule that applies prohibits a pilot from leaving the minimum altitude shown on the instrument approach chart (the minimum descent altitude for approaches without glidepath guidance or the decision height for full instrument landing systems) until and unless the runway itself or the runway or approach lights are in sight and the airplane is in a position from which a normal landing can be made. Strict adherence to that rule would take care of that type of accident, but the weather doesn't cooperate very well. The common landing minimums for a routine full ILS approach call for a decision height of 200 feet above the ground and visibility on the surface of ½ mile. (Lower minimums available for specially equipped airplanes and aircrews are used primarily by airlines; those lower minimums can go almost to zero-zero.)

The ILS is a good deal. If a pilot can fly the airplane precisely, or if the airplane has an autopilot that can fly it precisely, the airplane reaches an altitude of 200 feet above the surface right on track to continue to the runway. At that point, if everything is stable and the pilot just continues doing what he has been doing, the airplane will fly onto the runway. It takes a good hand on the controls, make no mistake about that, but it's still well-orchestrated guidance to the runway.

For a nonprecision approach, where there is no glidepath guidance, a typical minimum descent altitude might be 500 feet above the ground and the required visibility for landing is usually one mile. Theoretically, this would leave the airplane flying at an altitude of 500 feet, with the airport just coming into view a mile away. The pilot must then fly visually to the runway.

The weather doesn't tend to favor a nonprecision instrument approach. In going through records, I found that over a period of a few years I flew 106 instrument approaches. For half of those, the ceiling was 500 feet or less, meaning that the weather would have been too poor for the average nonprecision approach. In 23 approaches the ceiling was 600 feet, which you might say is marginal for a 500-foot minimum descent altitude because cloud bases can vary. So in almost three-quarters of the approaches the weather was either marginal or below the nonprecision approach minimums. Such approaches can still be safely attempted, but only if the letter of the law is followed about not leaving the minimum altitude unless the runway is in sight. The strongest temptation to descend lower comes at night, when lights can be seen directly beneath as the airplane moves through the lower part of the cloud. The thought is that the lower you fly, the more you can see. That simply isn't always true.

## WEATHER MINIMUMS

A couple of regional airline accidents in the late '70s, actually within a few weeks of each other, can be used to learn about this phenomenon.

The captain of the first flight we'll discuss contacted his company's weather observer at the destination airport and was told that the weather there was 700-foot ceiling and three miles' visibility. Once en route, the pilot was given destination weather of an indefinite ceiling at 300 feet with ¾ of a mile visibility and a notation that conditions were deteriorating. The destination didn't have a full instrument landing system. The weather at a nearby military base was much better. The flight was cleared for an approach to the original destination and as the aircraft was descending and maneuvering for an approach, the crew radioed that it appeared they would miss the approach and might need clearance on to an alternate airport. The aircraft crashed 1.2 miles before reaching the airport.

In the investigation of this accident, a number of alleged unsafe practices by the airline involved were explored. In a brief summary of these alleged unsafe practices, the NTSB included "company minimums" between 200 and 350 feet, which was below the FAA minimum for that approach. It was also alleged that unapproved instrument approach procedures were used, that they ignored takeoff and landing visibility minimums and directed pilots to make repeated instrument approaches and to "get lower" during adverse weather conditions. There were other allegations but these are the ones that relate primarily to the question at hand.

The NTSB concluded, among other things, that the flight descended below the MDA of 440 feet without the crew having visual contact with the runway environment and that there were company pressures to make every attempt to return the aircraft, even if it meant a descent to a lower altitude than approved minimums. Add below-minimum weather to pressure to complete the approach and you have the lethal combination.

In the other accident, which also occurred at night, the reported weather at the airport showed the ceiling to be variable between 200 and 400 feet with two miles' visibility in snow. The minimum descent altitude for the approach in use was 659 feet above the ground. The flight was cleared for an approach to the airport. Then the flight reported the approach and runway lights in sight at a time when the airplane would have still been well above the MDA and a number of miles from the airport. A bit later, the flightcrew requested that the approach lights be dimmed. The aircraft then struck the approach lights serving the runway in use just 300 feet from the end of that runway. Subsequently it crashed into an embankment about 200 feet from the runway.

During the course of the investigation, several former pilots of the airline testified that the MDAs for the published approaches at the airport were not adhered to. They stated that the company vice president for operations cleared individual captains for "company" minimums after he was satisfied that the captain was capable of flying the aircraft to lower MDAs. The "company minimums" involved MDAs of about 200 feet above the airport elevation, and the approaches were flown with the aid of distance-measuring equipment, to establish a position in relation to the airport. The purpose of the lower MDAs was to achieve a higher completion factor for the flights, since the officially reported weather conditions otherwise frequently precluded a successful approach and landing. The NTSB couldn't determine the reason for the pilot's premature descent but said, "It was probably the result of: (1) a deliberate descent below the published minimum descent altitude to establish reference with the approach lights and make the landing, (2) a visual impairment or optical illusion created by the runway/approach lighting systems, and (3) downdrafts near the end of the runway."

Both those accidents happened some years ago, when the commuter/regional airlines were more in the "fly-by-the-seat-of-your-pants" mode than they are today.

### HUMAN FACTOR

I would hope that no airlines pressure pilots to complete flights regardless of conditions today. But even with no pressure from management, the human factor is still there. Pilots are like anybody else. They want to complete the job they start and they like to get home on time. Fortunately, most of the time the airline crews have the equipment to safely fly to lower minimums and they have weather avoidance radar to stay out of storms, but things are not always ideal. My friends who fly the British Airways Concorde flights into New York are amazed that at JFK, the premier U.S. international airport, they are often called upon to make a nonprecision approach in their supersonic transport.

### THUNDER AND LIGHTNING

The FAA can make rules about weather minimums, and when there is a related accident it's usually possible to determine that the pilot did something that almost had to be contrary to the rules. Thunderstorms are a different matter. Through the end of 1985 there had been four big thunderstorm-related accidents in the recent past—an Eastern 727 at JFK, a Pan Am 727 at New Orleans, a Delta L-1011 at Dallas/Fort Worth, and a Southern Airways (now Republic) DC-9 near Atlanta—and several thunderstorm-related accidents that were less spectacular than those.

To a nonpilot observer, it might seem negligent of the FAA not to just make a law against flying through thunderstorms. Given the nature of the rules system, many would think this had already been done because that is the usual response: make a rule that would have prevented the accident. But in this case, it would not work for several reasons.

Who would decide when a thunderstorm exists? They come in all sizes, shapes, and intensities, and where some have little effect on surface weather and thus on the airport environment (where most thunderstorm-related accidents happen) other thunderstorms can turn the air around the airport into a crazy jumble of wind shear. If airports were closed by the FAA whenever a thunderstorm was reported (which is whenever thunder can be heard at the airport), there would be many days when the airline system would be thrown into schedule chaos. If air traffic control or National Weather Service radar were used, and a meteorologist were to make the decision, that might be better, but there might still be times when an airport would be closed even though safe operations could continue. Thunderstorms build, mature, and dissipate relatively quickly, and thus defy packaging by regulation.

Dealing with them has been exclusively the responsibility of the pilot-in-command and it is likely to remain that way.

The airlines seldom lose an airplane to a thunderstorm during the en route portion of a trip. The last one lost was the Southern Airways DC-9, near Atlanta, in 1977. This aircraft penetrated a severe thunderstorm, suffered a flame-out of both engines because of rain ingestion and hail damage, and while the crew made a successful power-off touchdown on a country road, the airplane was destroyed as it hit a gas station, trees, and other obstacles after touchdown. There were some survivors, but most were lost.

There are several reasons airplanes are lost to storms around airports instead of en route. All airlines have company procedures that tell crews not to fly airplanes in thunderstorms. And crews follow these procedures because the inside of a thunderstorm is an uncomfortable place to be, whether you're a pilot or a passenger in an airplane. No pilot would enjoy facing a cabin full of passengers after giving them a ride through a storm. And in the history of large jet transports few have suffered structural damage from thunderstorms, so that's not the problem. It might make passengers ill, scare them, and result in spilled coffee and even injury to anyone who isn't belted down. But the airplane will generally take the punishment of turbulence if a pilot gets too close to a thunderstorm while climbing, cruising, or descending. But the history of thunderstorm accidents suggests that pilots think differently about them in the vicinity of the airport. While you can fly some extra miles and avoid a storm en route, if one is menacing the airport there are but two choices. Either deal with the thunderstorm on the approach and landing, or go somewhere else.

At this writing, the National Transportation Safety Board had released the transcript of the cockpit voice recorder in the Delta L-1011 that crashed at Dallas/Fort Worth but not the full accident report. There are two places in this transcript that shed light on how pilots feel about thunderstorms. These are offered here only as an illustration. Any speculation on their relationship to the cause of the accident is not intended.

First, the following excerpts of conversation on the recording ensued as Delta Flight 191, the airplane that crashed, was descending into the Dallas/Fort Worth area. There were other airplanes in the area and communications with them are included just to give an example of how pilots were dealing with thunderstorms that were apparently rather numerous in the area. Intracockpit conversation between crewmembers is in italics.

DELTA 963: Delta 963, I'd like to deviate to the south.

CENTER CONTROLLER: Plane wanting deviation south, Delta 963, turn right heading of 260 to intercept the Blue Ridge 011 radial inbound.

DELTA 963: We're not going to be able to do that, sir, that's right in the middle of a big thunderstorm up here and we need to either stay on present heading or deviate slightly south and east of course.

CENTER CONTROLLER: Delta 963, I got an area 12 miles wide, all the aircraft are going through there, good ride, I'll have a turn back in before you get to the weather.

DELTA 191 FIRST OFFICER: *It would be nice if we could deviate to the south of 250.*

DELTA 191 CAPTAIN: *Somebody just ahead tried to and they wouldn't let them do it . . . they're working a 12-mile corridor. . . . The airplanes that have been going through there have been all right.*

CENTER CONTROLLER: 191 descend and maintain 10,000, the altimeter 29.91, and suggest now a heading of 250, 250 to join the Blue Ridge 011 radial and inbound, we have a good area there to go through.

DELTA 191 CAPTAIN: *Well. I'm looking at a cell at about heading of, ah, 255 and it's a pretty good size cell and I'd rather not go through it. I'd rather go around it one way or the other.*

CENTER CONTROLLER: I can't take you south. I got a line of departures

to the south. I've had about 60 aircraft go through this area out here, 10 to 12 miles wide there getting a good ride, no problems.
DELTA 191 CAPTAIN; Well I see a cell now about heading 240.
CENTER CONTROLLER: Okay, head . . . when I can I'll turn you into Blue Ridge, it'll be about the 010 radial.

Later, after the aircraft had passed that particular area of weather, the following:

DELTA 191 CAPTAIN: *You're in good shape. I'm glad we didn't have to go through that mess. I sure thought he was going to send us through it.*

During the descent, the crew was apparently concerned about avoiding the weather, as were other crews in the area. They might have flown through it if the controller had no other airspace for them to use, or they might have requested a diversion to another airport or area. It's very seldom that an air traffic controller will have to insist that a pilot fly into weather that the pilot doesn't want to fly into, and, if it looks bad enough, it's the pilot's pre-rogative to decline on the basis that it would compromise the safety of the flight. Then the controller would have to divert the aircraft to another area or move other airplanes around, and the pilot would have to answer for his actions after he landed.

In some major terminal areas, notably New York, the airspace is quite restricted. Traffic for three major airline and two major general aviation airports (La Guardia, JFK, Newark, Teterboro, and White Plains) in close proximity has to be handled. There is little available room for pilots to use in deviating around storms. You either stay on track or it doesn't work. So it is necessary to, in effect, close the area to inbound traffic and restrict outbound traffic when widespread thunderstorm activity moves into the airspace. When that is done the consideration is not directly related to airplanes landing or taking off in the vicinity of thunderstorms. It's just that in New York you couldn't handle all the airplanes going to five extremely busy airports through a 12-mile-wide corridor, as the controller was doing in the Dallas/Fort Worth situation just used as an illustration. To channel them through such an area and then channel them to individual airports would be impossible. Also, if an unbroken string of airplanes was inbound to JFK, for example, and that airport came under the influence of a severe storm, the controllers would have neither airspace nor other airport capacity to use in diverting those aircraft.

The FAA does a good job of anticipating thunderstorm problems with a central flow control facility that monitors areas of menacing weather and comes up with a strategic plan to reroute airplanes around bad areas or restrict operations into and out of busy terminals. That's the big strategic plan. Relatively isolated areas, where storms ebb and flow in a terminal area, come down to a much more individual operation. It's rather like the difference between a general coming up with the concept of how to maneuver his army, and the individual soldier about to engage in hand-to-hand combat with an adversary.

Back to Dallas/Fort Worth on August 2, 1985, and Delta Flight 191, to see more about how this might work. The airplane has completed its descent into the terminal area and is now in contact with Dallas/Fort Worth approach control. Again, cockpit conversation between crewmembers is in italics. The first officer was flying the aircraft.

DELTA 191 FIRST OFFICER: *We're gonna get our airplane washed.*
DELTA 191 CAPTAIN: *What?*
DELTA 191 FIRST OFFICER: *We're gonna get our airplane washed.*
DELTA 191 CAPTAIN: Approach, Delta 191 with ya at five.
APPROACH CONTROL: 191 heavy, expect 17 left.

DELTA 191 FIRST OFFICER: Thank you, sir.

APPROACH CONTROL: Delta 191 heavy, fly heading of 350.

DELTA 191 CAPTAIN: Roger.

APPROACH CONTROL: American 351, do you see the runway yet?

AMERICAN 351: As soon as we break out of this rain shower we will.

APPROACH CONTROL: Okay 351, you're four from the marker, join the localizer at or above 2,300, cleared for ILS 17 left approach.

AMERICAN 351: Cleared for the ILS, American 351.

APPROACH CONTROL: 191 heavy, reduce speed 170, turn left 270.

DELTA 191 CAPTAIN: Roger.

APPROACH CONTROL: Five Juliet Foxtrot, turn left 190.

5JF: Left turn 190.

APPROACH CONTROL: Five Juliet Foxtrot, increase your speed to 170 knots, hold that to the marker, you're five miles from the marker, join the localizer at or above 3,000, cleared for an ILS 17 left approach.

5JF: Cleared for the 17 left approach, roger, we're around to 190.

APPROACH CONTROL: Delta 191 heavy, turn left to 240, descend and maintain 3,000.

DELTA 191 CAPTAIN: 191, 240, outta five for three.

APPROACH CONTROL: American 351, tower 126.55.

AMERICAN 351: So long.

APPROACH CONTROL: November five Juliet Foxtrot is four miles from the marker, maintain a speed of 170 or better to the marker, you're cleared ILS 17 left, contact tower 126.55.

5JF: 126.95, good day.

APPROACH CONTROL: That's 126.55.

5JF: 26.55, good day.

APPROACH CONTROL: Delta 191 heavy is six miles from the marker, turn left heading 180, join the localizer at or above 2,300, cleared for the ILS 17 left approach.

DELTA 191 CAPTAIN: Delta 191, roger all that, appreciate it.

APPROACH CONTROL: Delta 191 heavy, reduce your speed to 160 please.

DELTA 191 CAPTAIN: Be glad to.

DELTA 191 CAPTAIN: *160.*

DELTA 191 FIRST OFFICER: *All right.*

DELTA 191 CAPTAIN: *Localizer and glideslope captured . . . 160 is your speed.*

APPROACH CONTROL: And we're getting some variable winds out there due to a sh-shower on short out there north end of D/FW.

UNIDENTIFIED DELTA 191 CREWMEMBER: *Stuff is moving in. . . .*

DELTA 191 CAPTAIN: *160's the speed.*

APPROACH CONTROL: Delta 191 heavy, reduce speed to 150, contact tower 126.55.

DELTA 191 CAPTAIN: 126.55, you have a nice day, we appreciate the help.

DELTA 191 CAPTAIN: Tower, Delta 191 heavy, out here in the rain, feels good.

TOWER: Delta 191 heavy, regional tower, 17 left, cleared to land, wind 090 at five, gusts to 15.

DELTA 191 CAPTAIN: Thank you, sir.

TOWER: American 351, if you can make that next high speed there, pull up behind Delta and hold short of 17 right this frequency. [*This transmission was to the American flight that had just landed on 17 left.*]

AMERICAN 351: 351.

DELTA FIRST OFFICER: *Lightning coming out of that one.*

DELTA 191 CAPTAIN: *What?*

DELTA 191 FIRST OFFICER: *Lightning coming out of that one.*

DELTA 191 CAPTAIN: *Where?*

DELTA 191 FIRST OFFICER: *Right ahead of us.*

DELTA 191 FLIGHT ENGINEER: *You get good legs, don't ya.*

DELTA 191 CAPTAIN: *A thousand feet . . . seven sixty-two in the baro . . . I'll call 'em out for you.*

DELTA 191 FIRST OFFICER: *Aw right.*

DELTA 191 CAPTAIN: *Watch your speed . . . you're going to lose it all a sudden, there it is . . . push it up, push it way up . . . way up . . . way up . . . that's it . . . hang on to the @#$%.*

GROUND PROXIMITY WARNING SYSTEM: Whoop, whoop, pull up.
TOWER: November 15 Juliet Foxtrot can you make the, ah, we'll ex-
pedite down to the, ah, taxi 31, and a right turn off the traffic's a
mile final. [*This to the business jet that had just landed.*]

The ground proximity warning system on the Delta jet continued sounding. One impact was recorded, followed by a second.
The tower's last instruction to Delta 191 was for the aircraft to
go around but the crash sequence had already begun.

Where the crew of Delta 191, and other crews, had wanted
to fly around the thunderstorm activity they encountered approaching the Dallas/Fort Worth area, neither they nor the preceding crews or those flying aircraft that were following expressed
doubts to the controllers about flying through the thunderstorm
that was on final approach. The American flight and general
aviation aircraft preceding Delta 191 made no immediate report
of trouble on the final approach.

**THE CAPTAIN'S DECISION**

In many accidents like this, one of the probable causes is
found to be the captain's decision to fly through the weather
ahead. If there is lightning in it, as there apparently was from
the first officer's comment, it's a real thunderstorm, so there
would be little doubt that there would be turbulence and wind
shear ahead. But the airplanes ahead flew through without com-
plaining. The one immediately ahead of the Delta flight was a
small business jet: if it could go through, surely a heavy L-1011
trijet could fly through it. Wind shear does different things to
different airplanes, though, as will be explored in a separate
chapter on that subject. Too, the thunderstorm could well have
intensified just as Delta 191 reached it. Thunderstorms can and
do change from moment to moment, and where one airplane

might find the passage acceptable, the next one in the barrel might find conditions to be completely unflyable, and not be able to keep from sinking into the ground short of the runway.

I was doing some research on wind shear with British Airways and, as pilots do, we fell to discussing some of the wind shear accidents. One of the captains at the lunch table wondered aloud how many pilots, after it became apparent that there was convective activity on final, concentrate on flying the aircraft through the bad stuff and on to a landing, and how many, at the first sign of wind shear, decide to abandon the approach. They were very much in favor of abandoning the approach at the first sign of strong wind shear and later I got to fly some wind shear models in their Boeing 757 simulator. The effect of the shear on an airplane is always impressive and if it is encountered on approach, the only salvation is to abort the approach and try to transition the airplane to a normal climb configuration. But if this isn't done promptly, it might not work, and it's not likely to ever lend itself to regulation.

As a US Air DC-9 was approaching Detroit Metropolitan Airport in 1984, there was obviously thunderstorm activity in the area. It was visible on radar and the following pertinent excerpts are from the cockpit voice recorder transcript, starting when the aircraft was about five miles from the airport, inbound on the instrument landing system. (The words in italics represent intra-cockpit conversation.)

US AIR 183 CAPTAIN: *Smell the rain . . . smell it?*
US AIR 183 FIRST OFFICER: *Yup, got lightning in it too.*
TOWER: Winds are 320 at 26, peak gusts 36, north boundary winds 270 at 16, east boundary winds 310 at eight, south boundary winds 290 at 22.
US AIR 183 CAPTAIN: *It's going to get choppier than @#$% here in a minute.*

US AIR 183 FIRST OFFICER: *Sure is.*
US AIR 183 CAPTAIN: *I'll betcha it is, it's right on the end of the @#$% runway.*
US AIR 183 FIRST OFFICER: *Tower, US Air 183 is with you for the right side.*
TOWER: US Air 183, Metro tower, 21 right, cleared to land.
US AIR 183 FIRST OFFICER: *Cleared to land, US Air 183.*
US AIR 183 CAPTAIN: *Full flaps.*
US AIR 183 FIRST OFFICER: *Out of a thousand.*
US AIR 183 CAPTAIN: *Wipers on.*
TOWER: Wind check, centerfield wind 320 at 27, east boundary wind 320 at niner, the north boundary wind 260 at 13.
US AIR 183 FIRST OFFICER: *Approach lights in sight.*
US AIR 183 CAPTAIN: *Rick, you might have to stand by these spoilers when we get on the ground.*
US AIR 183 FIRST OFFICER: *I'll stay right on top of 'em . . . okay . . . outta 1,500, speed is plus 15, 15 . . . sinkin' eight, no flags.*
FRONTIER 214: 214's going around.
TOWER: Frontier 214 fly runway heading, climb and maintain 3,000.
FRONTIER 214: Frontier 214, runway heading, up to three, roger.

Next the tower gave another wind report to a Northwest flight landing on the other runway, 21 left. The wind was given as "320 at 28, peak gusts 40, correction 42."

US AIR 183 CAPTAIN: *Ask 'em if they got the runway lights on, I can't get a word in edgewise.*
US AIR FIRST OFFICER: *All right.*
[*Sound of hail begins overriding all other audio.*]
US AIR 183 FIRST OFFICER: *Tower, US Air 183, turn the runway lights on to the left side please.*
[*Sound of hail subsides, sound of ground proximity warning horn begins.*]
US AIR 183 FIRST OFFICER: *US Air 183 missed approach.*
US AIR 183 CAPTAIN: *Down the gear, down the gear . . . down the gear . . . down.*

This was followed by the sound of impact. The captain later stated that as soon as the airplane entered the clouds, rain, and hail, he lost sight of the runway environment and immediately started a missed approach. The crew recalled that shortly after starting the missed approach climb, the rate of climb slowed and then stopped and the speed started to decrease; as the airspeed decreased the airplane started to descend. As it neared the end of the runway, the airplane flew out of the rain and hail and the captain saw the runway. He stated that he believed the airplane was still descending and that the airspeed had decreased more. Believing that ground contact was imminent, the captain ordered the landing gear extended. (It had been retracted for the missed approach.) After ordering the gear extended the captain pushed the nose over to assure a level touchdown and he pulled the thrust levers back to reduce power. The airplane touched down before the landing gear was fully extended and skidded 3,800 feet down the runway before stopping. There were no serious injuries.

The flight attendant and several passengers told National Transportation Safety Board investigators that the flight was routine until the airplane suddenly encountered severe turbulence, rain, and hail. They reported that they were shaken vertically and laterally during the approach though the belts were fastened snugly.

In this case, some other airplanes landed, but other pilots opted out of the approach. If there had been a rule against making approaches in a thunderstorm, and if the controllers or a meteorologist had the authority to shut an airport down, would this approach have been prohibited? Would the Delta L-1011 at Dallas/Fort Worth have been told to go away? There is probably no answer to either question because of the capricious nature of thunderstorms. Only one thing is certain. When approaches (or takeoffs) are conducted when thunderstorms are in the area, the risk increases rather dramatically. But the assessment of that risk is probably still best left to the pilot-in-command.

## Airlines Aren't Equal

For those riding airlines, the rules change a lot over the spectrum of airlines. It's interesting that the rules change as much as they do over operations that are called "airlines" and the changes give an insight into the thinking of the folks who have molded the air transportation system. There seems a proclivity to allow more risk to exist in smaller airplanes, presumably to keep the cost down and allow the development of small airlines with a minimum of regulation.

There are virtually no small airlines still operating single-engine airplanes, but if there were, this would be the minimum basis of regulation. The first step up is when the airline operates a multi-engine airplane weighing less than 12,500 pounds. Then the pilot of the aircraft is required to have an airline transport pilot certificate. To be eligible for this certificate a pilot must have 1,500 hours of flying time, meet other minimum flight time requirements, and be 23 years old. If the airplane has more than 10 passenger seats, the airplane must be operated by two pilots and must have thunderstorm detection equipment. If the airplane weighs over 12,500 pounds, the pilot-in-command must have a rating to fly the specific type of airplane. The aircraft must also meet takeoff performance requirements and certain fire protection standards. If there are over 19 passenger seats the aircraft must have weather avoidance radar and a flight attendant. With 31 or more seats the whole operation comes under the same rules as the major airlines.

In nonairline operations, the FAA's rules seem onerous to many pilots but in reality they are quite lax—more lax than in any other country in the world. The rules have been made as they are on purpose. The total freedom to fly as you please is unique to the U.S. and to Canada (though the rules are stricter in Canada) and it is a pretty wonderful thing to be able to do. The main requirement for the pilot, and for the airline passenger, to realize is that everything that's legal isn't necessarily free of risk. The FAA has a neat habit of always saying that everything is okay even though an airplane and its passengers lie in ruins, and the FAA has another neat habit of blaming everything on someone else. And again, in that rule vesting the ultimate authority in the pilot-in-command, the FAA indeed has a shield to use in fighting off the spears. When flying, remember this: if in a general aviation airplane, it's generally 100 percent in the hands of the pilot and the accidents that happen are on flights that are FAA approved and usually legal up to the last minute. When riding on an airline, the pilot does have the final authority but it is the airline (not the FAA) that sets the tone and the framework within which the pilot operates. So when buying tickets, the corporate personality of the airline becomes important to the passenger. And many observers feel that the trend toward low fares and hustle-up operations at hubs and spokes, combined with the increasing age of the airline fleets and the decreasing experience level of aircrews, will lead to a wholesale overhaul of some of the rules governing airline operations. But it will, naturally, only come if dictated by specific accidents.

# 4

## Pilot Error

FOR YEARS, pilot error has been cited as the cause of most airplane crashes. This is logical. If the pilot is the final authority, and if the airplane won't fly without the pilot taking all the necessary actions, then blame for most any crash can be laid at the feet of the pilot. But the subject is deeper than this. Some pretty fine pilots have been blamed for crashes; on the other hand, some crashes have involved blatant and reckless errors. It's unfortunate that they have to be all lumped together. A pilot faced with a quickly developing problem has to make split-second decisions with the continuous 50-50 chance of making the wrong decision. One slip can result in an accident. On the other hand, some pilots methodically fly past warning signals and into predictably impossible situations. Some pilots come to grief simply because they don't know better. Others develop a cavalier attitude about flying—they don't have or they lose that good old feeling that the airplane will kill them if they don't keep a constant guard. A pilot's error might be motivated by other factors—the press of business or a desire to get home are examples.

## Commuter Pressure

A complex but interesting example of all this is found in the report of an accident that had the following probable cause: "The National Transportation Safety Board determines that the probable cause of the accident was the failure of the pilot to execute the emergency engine-out procedure properly shortly after takeoff following a loss of power in the left engine because of water in the aircraft's fuel system and the failure of the Puerto Rico Ports Authority to remove excess water known to be in the airport's in-ground fuel tank before conducting fuelling operations. The pilot's failure to execute the engine-out procedure properly was due to his inexperience in multiengine airplanes.

"Contributing to the accident were: (1) the air carrier's use of a pilot not certificated for the flight; (2) the air carrier's failure to train the pilot adequately; (3) the pilot's failure to follow proper practices to detect water in the airplane's fuel tanks; (4) the out of weight and balance condition of the airplane; (5) the Federal Aviation Administration's incorrect application of 14 CFR Part 135 rules to commuter air carriers; and (6) the FAA's generally inadequate surveillance of the air carrier."

There's a lot there, but number one on the list was the pilot, a 21-year-old who had gotten his commerical pilot certificate and multiengine and instrument ratings less than five months before the accident.

The flight, in a twin-engine Britten-Norman Islander, was originally to have been a scheduled commuter flight, which would have required a pilot with an airline transport pilot certificate. But that flight was cancelled and the accident flight was designated an extra section, which the airline counter agent considered an on-demand air taxi operation for which the pilot would have been qualified.

The pilot had flown the airplane earlier in the day, and when preparing for the accident flight he taxied it to the gas pump where about 30 gallons of fuel were added to each tank. No witnesses were found who saw the pilot check for water in the fuel tanks or drain the sumps after the fuelling operation was completed.

The aircraft departed from the ramp with eight passengers aboard, all of whom had reserved seats on and had purchased tickets for the regularly scheduled flight.

After takeoff, the airplane appeared to climb normally to about 200 feet. There is appeared to lose power. It gained about another 50 feet while in a nose-high attitude. Then the airplane descended and crashed into the ocean. All aboard died. Later, another aircraft operated by the airline was checked for water in the fuel; it was there. According to testimony, the airport fuel tank had been checked for water before the accident airplane was refuelled and 1½ inches of water was noted in the bottom of the tank. There was a substantial difference of opinion about when and how the water was removed from the tank but some testimony was to the effect that a large amount of "water and mud" was removed from the fuel supply at the gas pumps after the crash.

A weight and balance manifest, which reportedly was prepared by the airline counter agent prior to the departure of the flight, was provided to National Transportation Safety Board investigators after the accident. A second manifest was provided to NTSB investigators by FAA representatives at a later date. The details of this manifest were significantly different from the original one furnished. The NTSB, in its report, prepared a probable loading based on passenger weights established during autopsies and other factors. This estimate showed the airplane to be more than 10 percent over its maximum allowable takeoff weight. The center of gravity of the aircraft was five inches aft of the limit.

### FAA INVOLVEMENT

The FAA had previously approved the airline to use pilots who did not meet all the requirements (an airline transport pilot certificate) but this exemption had expired at the time of the accident. The designation of the flight as an on-demand operation was apparently with some government blessing. The NTSB report stated: "The fact that local FAA personnel viewed the accident flight as an on-demand flight (until corrected by FAA head-quarters) after the accident strongly suggests that . . . management's interpretation and application of the Part 135 rules was condoned by the FAA; however, there was no evidence to support this conclusion other than what actually happened on the day of the accident." Eleven months after the accident the FAA swung into action by issuing an emergency revocation of the airline's oper-ating certificate. The order stated that the airline's officers and employees knowingly prepared a false flight manifest for the ac-cident flight and presented the fraudulent manifest to NTSB investigators. Also cited was the use of an improperly certificated pilot, the operation of the airplane without compliance with flight manual weight and balance limitations, and careless and reckless operational behavior which endangered the lives and property of others.

The first item in the probable cause, the "failure of the pilot to execute the emergency engine-out procedure properly," is open to some question if the NTSB's weight estimate is correct. The published single-engine rate of climb of an Islander is 145 feet per minute on a standard day at sea level. That's just barely climbing. The temperature at the time of the accident was 15 degrees centigrade above standard, which would have seriously degraded performance. Add to that the overweight condition, and it is hardly likely that the airplane had the performance capability to do anything other than crash away from the airport following an engine failure.

True, the 21-year-old pilot's hands were on the controls when the airplane went down, but there sure was a lot more to it than that. This pilot was faced with an immediate and compelling problem, with an awful lot of contributing factors. The outcome wasn't good.

## Electrical Failure

An accident involving a Hawker Siddley 748 twin turboprop in scheduled airline service shows how the in-flight decision process works when the problem is not of a split-second nature. The bottom line on this one was, according to the NTSB, ". . . the captain's decision to continue the flight toward the more distant destination airport after the loss of DC electrical power from both airplane generators instead of returning to the nearby departure airport. The captain's decision was adversely affected by self-imposed psychological factors which led him to assess inadequately the airplane's battery endurance after the loss of generator power and the magnitude of the risks involved in continuing to the destination airport. Contributing to the accident was the airline management's failure to provide and the FAA's failure to assure an adequate recurrent flightcrew training program which contributed to the captain's inability to assess properly the battery endurance of the airplane before making the decision to continue, and led to the inability of the captain and the first officer to cope promptly and correctly with the airplane's electrical malfunction." (One NTSB member disagreed with the inclusion of anything other than the captain's decision to continue to the destination.)

It was dark when the HS 748 left Springfield, Illinois, bound for Carbondale, Illinois. Right after takeoff, the flight called Springfield departure control and reported a slight electrical problem. Despite this, the plan remained to fly to the destination, about 40 minutes' flying time away.

The generator on the left engine of the aircraft had failed and in responding to the failure the first officer apparently, and mistakenly, isolated the right generator and bus bar from the aircraft DC electrical system and, thereafter, the right generator disconnected from the right generator bus bar. Attempts to get the right generator back on line were unsuccessful and the airplane flew on, relying solely on its nickel cadmium batteries for electrical power. The flight was conducted in clouds. There were scattered thunderstorms in the area and the visibility below the clouds was one mile in rain.

As they worked on the problem, the first officer told the captain that "the left (generator) is totally dead, the right (generator) is putting out voltage but I can't get a load on it." A couple of minutes later, the first officer told the captain that the battery power was going down "pretty fast."

Subsequently, the captain told the air traffic controller that he had an "unusual request." He wanted to descend to 2,000 feet, a safe altitude but lower than would normally be used. As the flight droned on through the night clouds, and about seven minutes after the electrical problem started, the captain began giving the first officer instructions to turn off lights. When the airplane was 40 or 50 miles away from the destination, the first officer told the captain, "I don't know if we have enough juice to get out of this." A descent was started, presumably to try to get below the clouds where ground lights could be seen. The captain told the first officer to have his flashlight ready, the first officer said they were "losing everything," and the captain asked the first officer if he had any instruments on his side. The airplane then descended, turned about 180 degrees, and crashed.

All the HS 748 flight instruments are operated by electrical power. In the event that both generators fail (or all generators fail, or there is a general failure in an electrical system), large aircraft are required to have either an alternate source of power (such as a ram air turbine) or provisions for emergency use of the batteries at a reduced load for at least 30 minutes of operation. The ground school instructor for the airline testified that pilots were taught that if both generators failed and could not be restored, the flightcrew should reduce the electrical load on the center bus bar to 35 amperes for each of the two pair of batteries (a total of 70 amperes) and that a minimum battery endurance of 30 minutes was available. There was also a procedure to follow in the aircraft that would have isolated only emergency lighting and the most basic instruments required for flight, the turn and slip indicators (one on each side), airspeed indicators, vertical speed indicators, and the altimeter.

The investigation of this accident was almost as loaded with factors as the previous one. One item thought significant was the fact that there was a continuous reference during the flight to the number of volts left in the electrical system. From the NTSB report: "However, one of the inherent characteristics of the nicad battery is the fact that it will maintain voltage very close to its rated output until the battery cells are nearly depleted. Thereafter the battery output voltage decreases at a very rapid nonlinear rate. . . ." This is a different characteristic than would be found with lead-acid batteries, used on many airplanes, but the captain of this flight had flown over 3,000 hours in the HS 748, so lack of knowledge on this subject couldn't be charged to inexperience in the airplane.

### WHY CONTINUE?

Even though the crew did not reduce the electrical load to an absolute minimum for emergency operation, the batteries powered the airplane systems for 31 minutes after the failure.

The primary question relates to the pilot's decision to continue to Carbondale instead of considering the failure of all generating power as enough of an emergency to dictate the quickest possible landing at the closest airport. "The board believes that the captain's decision to continue was based on his reluctance to remain overnight in Springfield, his self-imposed determination to adhere to schedule, his demonstrated willingness to assume what he believed to be reasonable risks to adhere to schedule, and, in this case, a misplaced confidence in his knowledge of the airplane and his flying capabilities. Based on these factors, the captain did not evaluate properly the risks involved in continuing to Carbondale, and the safety board concludes further that his decision to continue not only was imprudent, but was improper."

As a result of the accident, and of an FAA special investigation, the FAA implemented programs designed to enhance the efficiency of surveillance of airlines. More inspectors were hired and new training programs were developed.

## Taking Risks

All of us, in everything that we do, go through those moments when the devil on one shoulder urges us on in something that we feel might be questionable, while the voice of caution whispers into the other ear. In flying, this has special significance and relates directly to risk. There is always some risk in flying, risk that ebbs and flows with other factors. The risk is seldom precisely the same for any two flights and the relationship of mechanical problems to risk is a key factor in pilot motivation to make a good decision as opposed to an error. Airplanes are seldom in precisely perfect conditions—there's usually a little something out of order all the time. They are complex enough for that, but it's my job, as the pilot of my airplane, to sort out what's okay and what isn't.

For example, right now I know that there is some inoperative panel lighting on the aircraft but there is other lighting that can do the same job. I always carry three or four flashlights, and no regulation is violated by flying with these particular lights inoperative. Any risk that is added is managed by the options.

Electrical failure: the airplane I fly offers as good as or a better deal than the HS 748 after the total failure of generating capacity. Such an event would be (and has been) handled in one of two ways. On a pretty, clear day, it would be reasonable to shut down all electricity and fly a considerable distance to the most convenient place to get the airplane repaired, saving the 30 or more minutes of battery capacity for the landing. The other way would be at night, when flying in bad weather, or both. The drill then would have to be to go to the closest suitable airport and land. Flying in the U.S., there is almost always a suitable airport within 30 minutes' flying time. In the rare event one isn't available, then there's always the totally independent battery-powered radio that can be used for communications while flying to a suitable airport. There are times when going to an airport other than the destination can be terribly inconvenient, but airplanes are machines, machines malfunction, and when they do the pilot's job becomes one of not allowing the malfunction to compromise the safety of the operation.

## Simple Omission

The two examples offered so far have involved a lot of factors. There are occasional accidents that involve something very simple, where the slightest omission leads to a major event.

An airline Boeing 727 was being flown on a ferry flight to pick up a load of passengers at another airport. The aircraft was cleared to climb to Flight Level 310 (31,000 feet). As the airplane was climbing, the recorded airspeed started to increase (the recording would show the same as that displayed to the pilots) and in response to this increase in indicated airspeed the pilots pulled the nose of the aircraft up to control airspeed. At 23,000 feet the overspeed warning sounded at a recorded airspeed of 405 knots. The sound of the stall warning stick shaker was recorded intermittently after the onset of the overspeed warning. A buffeting of the airframe was identified by the crews as the Mach buffet, which occurs when the airplane exceeds its critical Mach number. The captain then commanded the first officer, who was flying, to "pull it up." Next the recorded altitude of the aircraft started decreasing at a rate of 15,000 feet per minute and the aircraft started turning rapidly to the right. The crew transmitted an emergency message, reported that they were in a stall, and worked to recover from the stalled condition. The aircraft continued down and reached the terrain, 1,090 feet above sea level, 83 seconds after it reached its highest altitude, 24,800 feet.

The NTSB's probable cause was "loss of control of the aircraft because the flightcrew failed to recognize and correct the aircraft's high-angle-of-attack, low-speed stall and its descending spiral. The stall was precipitated by the flightcrew's improper reaction to erroneous airspeed and Mach indications which had resulted from a blockage of the pitot heads by atmospheric icing. Contrary to standard operational procedures, the flightcrew had not activated the pitot head heaters."

That was simple. And while one error, failing to turn the pitot heaters on, should not result in a disaster, there were extenuating circumstances.

When the pitot heads freeze over the effect on the airspeed indication can be to make it increase as the airplane climbs, as apparently happened in this case. So what the pilots saw on the instrument panel was the indication of increasing airspeed as the airplane was climbing. Why did they think that the airplane could do this? The safety board recognized the absence of comments about possible instrument error or airspeed system icing and concluded that the flightcrew attributed the high airspeed and high rate of climb to the aircraft's relatively low gross weight and to an encounter with unusual weather, which included strong updrafts. The safety board recognized that the flightcrew's analysis must have been influenced by the fact that the airspeed indicators on both sides of the panel were indicating the same. But the NTSB also said that the aircraft's attitude should have warned that the performance was abnormal because the nose was about 25 degrees higher than required for a normal climb and that at such a high nose-up attitude it would have been impossible for the airspeed to continue to increase even if influenced by a strong updraft. But that simple act of not turning on the pitot system heat was apparently what started the whole sequence.

## The Big Variation

Those three examples of accidents where the NTSB found the pilot or the crew to be primary in the probable cause shows quite a variation in the things that might result in such a conclusion. In the first two there were a lot of other factors; in the third, the 727, one simple mistake was apparently the catalyst.

In general aviation flying, the factors are usually simpler. Weather, or a misjudgment of weather, is a big factor. Often where the first news reports of an accident will say that the engine spluttered or the pilot was seeking a place to land, the fact of the matter is that a pilot who was not trained in and rated for instrument flying was, in fact, flying in inclement weather. A lot of airplanes are lost this way each year and if the weather is bad during a three-day weekend or holiday period, the problem is aggravated.

This is being written at Christmastime 1985 and at this very moment the TV is alive with news reports about a twin-engine airplane crashing through the roof of a shopping mall in California. The TV reporter said the pilot was groping in fog to land at a nearby airport, and that one approach had been unsuccessful and the pilot was coming around to try another one.

The investigation of that accident is just starting so the facts are not known. There is, though, a long history of light airplane crashes in low-weather conditions, and a fair percentage of them involve the pilot missing one approach and coming back for another one. Unless the first approach was improperly conducted or there is evidence that the weather has changed substantially, it might be considered a basic error to try again. The minimum altitude to which it is safe to descend doesn't change from one approach to the next. If the runway or the approach lights aren't visible at that point, then things won't be any different in a few minutes. The only real hope for a second approach would be for the pilot to cheat and go below the minimum descent altitude or decision height. Airplanes are, and have always been, very unforgiving of pilots who fudge a few feet here and a few feet there. Eventually it will always add up to too many feet.

## Unhappy Holidays

Some years ago I spent a lot of time between Christmas and New Year's out looking for lost airplanes. A holiday brings out a strong tendency of pilots to put extra effort into getting there, wherever *there* may be, and disaster. This time was an unhappy example of how this doesn't work, especially when the pilot isn't qualified and current on instrument flying.

The weather was lousy the weekend after Christmas and a Cessna 206 failed to complete a Friday night flight from Springfield, Missouri, to Jackson, Mississippi. A Mooney did not arrive at Batesville, Arkansas, late Saturday, as planned, and a Cessna 182 disappeared Sunday while trying to reach Booneville, Arkansas.

At the time, I was a member of the Arkansas State Aeronautics Commission, an unpaid appointed group that looks after the specific interests of aviation in the state. It was Monday before we heard about lost airplanes. On the TV Monday morning the first report said one airplane was missing. Then two. Then three.

We had no official role in the search, that being the purview of the Civil Air Patrol and the Air Force, but did have an unofficial obligation. So I went, with director Eddie Holland, to the search headquarters on Monday morning. The weather was lousy—we had to fly instrument flight rules to get there—and no primary aerial search activities were being conducted. The CAP can search visually only in relatively good weather.

Because all aircraft are required to have an emergency locator transmitter that will activate automatically in case of crash, an electronic search was being conducted by a USAF C-130, flying fairly high at 12,000 feet. We went flying at 4,000 feet to see if we could hear anything. There were ground parties, but searching the hills of Arkansas on foot for something as small as a wrecked airplane, with no firm idea of its location, could only be considered as well-intentioned exercise. They could hear us flying over but we could hear no signals and they could see no downed airplanes.

On Tuesday a helicopter pilot found the Mooney. It had crashed in relatively flat country and all four on board were dead. The Cessna 182 was found on Wednesday, when the weather had improved enough for visual searching to be effective and after an ELT signal had helped pinpoint its location. Two dead. The Cessna 206 was found on Thursday. Southbound, it hit the first hill it came to in the Ozarks, about halfway up. Five dead.

Often you read in the papers that an airplane that crashed "was not on a flight plan." The implication is that with one you are found quickly, without one, you feed the buzzards for a long time. It didn't work that way in this case—the 206 was the first airplane down and the last airplane found and it was the only one on a flight plan. (The search for the others started when people waiting for the airplanes to arrive reported them missing.)

So when three different pilots took on the misty hills of Arkansas, none of them made it. All three accidents were classic cases of a pilot continuing visual flight into adverse weather and colliding with the terrain. There was a hint of alcohol involvement in one, which might be considered the ultimate pilot error.

## Again and Again

There's another factor here that I found a recurring theme when I lived in Arkansas. Everyone thinks that Arkansas is flat. A writer for *Flying* magazine once referred in an article to Arkansas as flat. Yet the western, northwestern, and northern sections of the state range from hilly to downright mountainous. This was such a problem that warnings were printed on charts and were broadcast over navigational aids to alert pilots to what they should have already known. A lot of airplanes, including military, airline, and general aviation, hit those hills—I've heard that as many as 60 have done it, though no precise verification of the number exists to my knowledge. The cockpit conversation between the crewmembers of an airliner casts light on what folks might be thinking as they misjudge the height of the Arkansas mountains.

The night flight, in 1973, was flown in a twin turboprop Convair airliner that was scheduled to operate between El Dorado and Texarkana, Arkansas, a distance of 61 miles. There was, however, a line of thunderstorms between the two cities. The captain of the flight elected not to use his instrument flight plan. Instead, he opted for visual flying. This was not totally unusual at the time in prop airliners flown over short distances at relatively low altitudes, and many of the old school to this day feel that the best way to deal with thunderstorms in flat country is to fly as low as possible. Night, though, does add to the degree of difficulty in doing this.

After the aircraft left El Dorado, the crew spoke to no one on the ground over the radio and when the aircraft failed to arrive at Texarkana, a massive search started. I was in Arkansas at this time as well, on the Aeronautics Commission, and joined the frustrating search. The weather was good, so no problem there, but we just couldn't find the airliner. Over 200 airplanes flew well over 1,000 hours; one search airplane crashed killing three, and the airliner wasn't located for three days—a record length of time for finding a lost airliner in recent domestic history. When the airplane was found, it was 91 miles north of Texarkana, about that distance off course since from El Dorado to Texarkana is basically an east to west flight.

After takeoff, the crew discussed the weather at length and the airplane was flown parallel to the line of storms, in a north-northwesterly direction instead of west toward Texarkana. They were looking for a gap in the line of storms through which they might fly without encountering hail, heavy rain, or turbulence. The airplane's radar was used to keep tabs on the weather. In reading the conversation between the captain and first officer, remember that this was several years ago and that informal exchanges between crewmembers during cruise flight is not unusual.

CAPTAIN: If we get up here anywhere near Hot Springs, we get in the mountains.
FIRST OFFICER: Uh, you reckon there is a ridge line along here somewhere? Go down 500 feet; you can see all kindsa lights. Let's go ahead and try for 2,500.
CAPTAIN: All right . . . you can quit worrying about the mountains 'cause that'll clear everything over there.

FIRST OFFICER: That's why I wanted to go to 2,500 feet. That's the Hot Springs highway right here, I think.

CAPTAIN: You 'bout right.

FIRST OFFICER: Texarkana . . . naw, it ain't either. Texarkana's back here.

CAPTAIN: Texarkana's back over here, somewhere.

FIRST OFFICER: Yeah, this ain't no Hot Springs highway.

There followed some discussion about the radar and the fact that as they flew along, thunderstorm after thunderstorm kept appearing. It was a long line. The first officer remarked that if they kept going they would wind up in Tulsa, but there was very apparently a healthy respect for thunderstorms and no desire to fly through them. There was discussion about the airplane being in clouds (which it was not supposed to be when flying by visual flight rules) and discussion of seeing ridges on the airborne radar.

FIRST OFFICER: Paintin' ridges and everything else, boss, and I'm not familiar with the terrain. We're staying in the clouds.

CAPTAIN: Yeah, I'd stay down. You're right in the base of the clouds. I tell you what, we're going to be able to turn here in a minute.

FIRST OFFICER: You want to go through there?

CAPTAIN: Yeah.

There was then discussion about the weather radar return and the headings the first officer should fly (he was flying the aircraft) as they turned toward the west to get through the line of weather at the best place. As the turn continued, the conversation tells the story.

FIRST OFFICER: Well, we must be somewhere in Oklahoma.

CAPTAIN: Doing all the good in the world.

FIRST OFFICER: Do you have any idea of what the frequency of the Paris VOR is?

CAPTAIN: Nope, don't really give a . . . Put, a, about 265, heading 265.

FIRST OFFICER: Heading 265. . . . I would say we . . . up.

CAPTAIN: Think so?

[*Laughter*]

CAPTAIN: Didn't we? . . . Descend to 2,000.

FIRST OFFICER: 2,000, coming down. . . . Here we are, we're not out of it.

CAPTAIN: Let's truck on . . . about five to the right. Shift over a little bit if you can. . . . Sure can, that's all right.

FIRST OFFICER: Right.

CAPTAIN: That's all right; you're doing all the good in the world. I thought we'd get, I thought it was moving that way on me, only we just turned a little bit while you was looking at the map.

FIRST OFFICER: Look.

CAPTAIN: First time I've ever made a mistake in my life.

FIRST OFFICER: I'll be . . . man, I wish I knew where we were, so we'd have some general idea of the terrain around this place.

CAPTAIN: I know what it is.

FIRST OFFICER: What?

CAPTAIN: That the highest point out here is about 1,200 feet.

FIRST OFFICER: That right?

CAPTAIN: The whole general area, and then we're not even where that is, I don't believe.

FIRST OFFICER: I'll tell you what, as long as we travel northwest instead of west, I still can't get Paris.

[*Whistle*]

CAPTAIN: Go ahead and look at it.

[*Whistle*]

FIRST OFFICER: 250, we're about to pass over Page VOR. You know where that is?

CAPTAIN: Yeah.

FIRST OFFICER: All right.

CAPTAIN: About 180 degrees to Texarkana.

FIRST OFFICER: About 152.
FIRST OFFICER: Minimum en route altitude here is forty-four hund—

In the middle of the first officer's sentence, the airplane, flying at about 2,000 feet, collided with the steep and wooded north slope of Black Fork Mountain about 600 feet below the top of the mountain. At the time, there was a warning on the Page VOR, "Caution, elevation 2,700 feet," but there was nothing on the recording to indicate that the crew had listened to the identification of the station and had thus heard the warning. Apparently they just tuned the frequency of the navaid without listening.

This was a much-discussed case. The pilots could fly the airplane very well, which is often not the case when small airplanes are lost in inclement weather. They were following a very conservative course in relation to the thunderstorm activity. Flying in the clouds without an instrument flight rules clearance is illegal but at night, in that area, with bad weather on the prowl, some pilots would rationalize it with the feeling that theirs is probably the only airplane within 50 miles. But mountains are where you find them, not where you think they are, and the flight came to a tragic ending primarily because of a lack of knowledge of the geography of that area.

## Monthly Treatment

In *Flying* magazine we had a column entitled "Pilot Error" that was renamed "Aftermath" to reflect the fact that *all* accidents are not a result of pilot error. Printed just about every month, it chronicles an accident. Done with the hope that it will help other pilots avoid tumbling a similar line of dominoes, it is one of the most popular columns in the magazine.

Originally, this was always based on a transcript of conversation between the pilot and a ground facility (in the case of general aviation aircraft which have no cockpit recorder) or of that combined with cockpit conversation. After it had been running for a while, we realized that there were a limited number of accident types, and that we had run about two of each. The transcripts were thus getting repetitive. The bright idea was born to flag them as fiction, make up the transcripts, and add the pilots' thoughts in italics.

These were fun to write because they were not about actual events, and because it was interesting to take a particular fictitious accident (that was based on a composite of real ones) and try to put together an image of what the pilot might have been thinking as he flew to fictitious oblivion. But this approach was not popular with the readers. Pilots read about accidents because they want to learn from the real mistakes of others. Imaginary pilot errors won't do. And while it's never a lot of fun to have an idea shot down by your readers, it was heartening to see the strong interest in learning what to avoid that was evident in this experience.

## Fine Line

As pilots study accidents to learn causes, it becomes clear that in many cases there is a very fine line between pilot error and other factors. A recently released and excellent FAA publication on winter operations gives at least one example of the old "you can and you can't; damned if you do, damned if you don't."

The paper starts off with a discussion of anti-icing additives for fuel. Because of the extreme low temperatures encountered at higher altitudes, water can form ice crystals in fuel and even though water is heavier than fuel and should settle to the bottom of the tank where it can be drained, these ice crystals can become suspended in the fuel. The ice can in turn block fuel filters, make fuel selector valves stick, or do other things that can lead to engine failure. Adding anti-icing additives to fuel can eliminate this hazard.

When out flying a new model of the airplane that I usually fly, I was given several cans of additive and was told to use it in the fuel of this airplane. I wondered why, having flown a similar airplane for thousands of hours without additives or problems. Then, when looking at the log book in the new airplane, I found out why: despite the fact that the airplane was just a pup, with few flying hours, pilots had recorded two cases of power interruption at very cold temperatures. The additive, trade named Prist, was the solution.

The pilot's operating handbook on my airplane had always mentioned the possibility of using the additive, and so I went in search of wise words on the subject. The person I asked was an aeronautical engineer who had worked with the specific type of airplane. He suggested that given the good operating experience with the airplane when not using the additive, I'd not be wise to start using it. There was a possibility, he said, that it would cause some things in the fuel system to start to leak.

So, on one hand the FAA says that it is a good idea to use additives and on the other hand an engineer, who has spent a lot more time thinking about this specific airplane than has the FAA, says not to use the additive. If the airplane is flown on without the additive and has an icing problem, is it "pilot error" for not following the FAA suggestion to use the additive?

That is strictly a judgment call, and lack of judgment or poor judgment is often cited as playing a role in accidents that are related to pilot error. Judgment, though, is perhaps not the best word to use here. At least a pilot sits in judgment of but one thing—his own performance both in the technique of flying the airplane and in the management of risks.

## *Quick*

There is a split-second nature to aviation that always has to be considered when evaluating the pilot's role in an accident.

When an airplane is flown on final approach for a landing, the reference speed is an important number. The runway length is a finite number, and when the length is calculated to be adequate for the landing in the specific airplane, it is based on the approach being flown at that reference speed.

Go now to a turbulent day with some wind shear. In such a condition a pilot would generally add a little extra speed to compensate for wind shear or for gusty conditions. For the sake of discussion, let's say that the pilot adds 10 knots of extra speed. The airspeed will vary as the airplane moves through gusty conditions, and when the airplane comes over the end of the runway, it might be flying at an airspeed higher than that which the pilot selected, or it might be a little slower. There is no way to tell with certainty which the case will be. Let's say that it's 15 knots over the original reference speed as the airplane passes over the runway threshold. Decision time! That extra 15 knots means that an airplane will require about 1,500 feet more runway to land and stop than it would otherwise have required. Is the extra runway available? What if that's exactly how much extra there is to spare? What if the wind is across the runway and is shifting in direction in a manner that adds a little tailwind component to the landing? These are all factors that have to be crunched in the ultimate aeronautical computer, the pilot's brain, and the airplane continues relentlessly on as the pilot contemplates the factors and decides whether or not the landing can be safely made on the runway given all the conditions.

You might say that all this should have been figured out in advance, and a pilot can and should make all the calculations in advance, but the final decision has to be made based on how it looks and feels at the last minute.

## *Parallel*

A medical researcher called our office and wanted a collection of reprints of our "Pilot Error" and "Aftermath" series. I asked him why he wanted these, and he said that in studying the interaction between physicians and other personnel in an operating room, he found a strong similarity between that and the interaction between flight deck crewmembers. Certainly in both places everyone must understand both what they are doing and what the other participants are doing. However, there is one big difference. In many cases, the physicians can stop, pause to think, or can mull the next move. There are, I'm sure, some cases where quick decisions and absolute accuracy are called for in the operating room, but this probably doesn't happen nearly as often as it does in the cockpit of an airplane. Every flight has a time past which it cannot go—that would be fuel exhaustion—and the decision process has to be completed successfully within that time.

I was on the flight deck of Concorde approaching New York,

moving along at perhaps 600 knots as the beautiful British Airways transport decelerated and descended for its landing. It's a very precise business of matching altitude and speed and rate of descent with the distance left to go, and the crew was doing a masterful job of making it all come out even. Then the person in charge of cabin service, who had some new help this trip, came forward and told the captain that he wasn't sure they would be able to get the cabin cleaned up in the time left before landing. The captain turned in his seat, gave the steward an icy stare, and said, "Well, what do you suggest I do, stop here and wait for you to get the bloody mess straight?" Needless to say, there was much scurrying about in the back and the cabin was put in order in the allotted time.

On one hand, the pilot has the pressure of getting everything done in the span of the flight and has the job of doing everything correctly. That's the flying technique and risk management area. On the other hand, the pilot has outside influences that increase or modify that pressure.

Confession is good for the soul and because this page is being written on January 1, I'll tell you a story that relates to this, a story of a flight that I made on New Year's Day of 1953.

## War Story

I was a hotshot charter pilot at the time, very junior in seniority at the Camden, Arkansas, airport where I worked. When a bunch of Georgia Tech fans wanted to go to New Orleans to the Sugar Bowl, to see their team play Ole Miss, I was selected to be captain of the ship, which was a World War II surplus twin-engine

Cessna, fondly dubbed the Bamboo Bomber because it was made of wood covered with fabric. A five seater, it was imposing enough but lacked anything but the most rudimentary radio and instrument equipment. The reason I was chosen to make the flight was, I think, that nobody else wanted to crawl out of bed on New Year's Day and go flying. I didn't do so well at this myself, and was a little late getting to the airport where my eager passengers waited.

The trip to New Orleans, about 300 miles away, with one stop en route to pick up another passenger, was relatively uneventful. The ballgame was very eventful for my passengers; Georgia Tech won by a score of 24 to 7. They were due back at the airport right after the game, but called in a somewhat celebratory mood and said they were going to get something to eat. The weather to the north wasn't getting any better and I spent the time before darkness furtively glancing in that direction, wondering how it would be to fly that old twin at night in reducing visibility. The thoughts were not happy ones, they were pretty dark, and when the passengers called back later and said they still had more places to go, I called time out. Tomorrow, I said; you folks keep on partying and we'll fly home tomorrow. Not only did I not relish the night trip, I didn't know whether my passengers would collapse into the airplane and sleep all the way back, or whether they would continue celebrating the Georgia Tech victory.

I called home base, and when I told the manager there that I wasn't coming back that evening, his tone of voice was such that I thought he thought I was chicken. Chickens don't fly well, you know. He asked me for a new estimated time of return and I told him we'd be back by noon the next day. Because of other pressing matters or duties for the airplane, I have forgotten which, he hinted that I had better be back by noon tomorrow. Pressure.

The next morning was overcast, one of those slate gray January days that seem so bleak. As best I recall, the passengers arrived on time, if a little tired from the festivities of the evening before. I also recall that the passenger we picked up on the way down had opted to go home on the bus. Whether his lack of confidence was in the Bamboo Bomber or its juvenile delinquent captain, I know not.

This was it. Get the airplane and the passengers home. By noon. The weather was marginal, and as the airplane lumbered off the runway at New Orleans Lakefront Airport and I herded it around toward the northwest, the view out the windshield was not encouraging.

With my high level of inexperience, I wasn't yet really in the risk management business. I knew something about what I was doing, but on this day I was flying with but one purpose—to get the airplane to the destination. I had yet to become a student of airplane accidents, so I wasn't really aware that what I was doing was the drum roll that precedes so many crashes.

Not long after the north shore of Lake Pontchartrain passed under my wooden wings, it became very difficult to maintain visual contact with the ground. That's the cornerstone of visual flight rules flying, and it wasn't working.

Decision time. I had some rudimentary training in flying on instruments, and the airplane had the most basic instruments required to fly in clouds. The decision was whether to climb up into the clouds and hopefully get on top of them, or turn around and go back to New Orleans. Pride, or buttheadedness, or pressure, or whatever you would like to call it, dictated that the airplane's nose would remain pointed toward the destination. Climb it would be.

At this point, the cloud layer was still relatively thin and the climb to an altitude that was on top of all clouds was neither lengthy nor eventful. (At that time, it was relatively legal to do this because an instrument flight rules clearance was not necessary to fly in clouds in much of the airspace around the U.S.) So I was on top of the clouds, feeling very much like an "all-weather" pilot, like an airline captain. But now the chore became navigating the airplane to the destination using the unsophisticated radio equipment in the old Cessna. In retrospect, my minute-to-minute decision making improved along in here because I made a good plan to fly to Monroe, Louisiana, using the old low frequency radio range there (on which you could fly along any of four legs, adjusting the heading of the airplane to keep it on the beam, where a constant tone was heard). From Monroe, I would fly the north leg of the range for a specified number of minutes, at which time I should be close to the destination. Then I'd descend through the clouds, find my exact position using a map, and navigate on home. There were but two problems that were putting a big spike in the risk of the day. One, the cloud tops had become higher which probably meant that the cloud layer was much thicker than that through which I had climbed earlier. There would be a longer period of instrument flying. Two, I couldn't raise anybody on the radio to get a weather report.

At the time the descent was started, I had a fleeting moment of strong misgiving. I remember thinking that because of the pressure to get the passengers back on time, I had gotten them, and myself, in a situation that I wasn't really qualified to handle. My short-term navigational decisions might have been okay but the long-term decision to fly the airplane into this predicament was awful.

When the moist clouds enveloped the Cessna, and it was a

matter of my looking at those basic instruments and properly controlling the airplane—or else—I felt very lonesome. I was working hard, and while the job that I was doing wasn't the smoothest in the world, the airplane was descending and proceeding out the north leg of the range with a reasonable degree of precision. It was here that I realized there were no options left. This was it. The clouds in which I was flying covered a wide area, and there was far from enough fuel in the tanks to fly to the nearest clear weather. That left but one choice. Descend until seeing something. The good news: this was in the days before TV towers, and the country was flat. The bad news: what if it was foggy? What if the clouds were right on the ground? There was still only one option. Descend gingerly until seeing something.

We were fortunate folks because there was enough ceiling and visibility beneath the clouds for me to fly in visual conditions, find a landmark, establish a position, and fly on to the home airport. When the passengers bade their captain a fond farewell, I kept up a good front but when they left my knees began to get weak. Mistakes, errors—I had made so many that day. It was pure luck that the weather was good enough under the clouds for me to fly. If I had wrecked the airplane, it would have been a classic case of "pilot error." The fact that I was young, inexperienced, under pressure, and didn't know any better would hardly have mattered.

A lot of pilots grumble about the fact that regardless of what happens, it's almost always found to be our fault. There are various levels of "pilot error," from simple omissions through outright flagrant careless operation, and perhaps pilots would be more comfortable if there were various levels assigned in accident reports. On the other hand, if we start laying blame on other factors, or on pressures, or on machine design even though nothing on the machine failed to work, then we aren't facing reality. The reason the regulations say the pilot-in-command is responsible for and the final authority as to the operation of the aircraft is that that's the way it works—and that is the only way it can work.

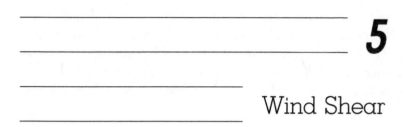

# 5

# Wind Shear

THE EFFECT of wind shear on airplanes has become a subject of almost mystical proportions. After any major accident people ask, "Was it wind shear?" Hardware is being developed to warn about wind shear, and all major airlines and large corporate flight departments are training in-flight simulators that are programmed with models of the types of wind shear that have caused famous accidents.

Wind shear has been around forever. It's really not anything new. In jest, one veteran airline pilot, now retired, said that he sure was glad he hung up his hat before wind shear was invented so that he wouldn't have to deal with it. Still, wind shear has become a larger factor in the accident picture.

What is wind shear? It is basically a change in wind direction or velocity that occurs with a change in height, or over a given distance. It's a changing wind. It's also thought of in connection with the downdraft of a thunderstorm. The best way to "see" wind shear is to watch how conditions develop as a thunderstorm approaches. When it's a distance away, the wind might be calm, or out of the south. As the storm approaches, there's what's called a gust front—air flowing out of the storm. The wind might shift from southerly to westerly and become much stronger as well as cooler. In the area where the wind is shifting, wind shear exists. The cooler air is coming down from aloft in the downdraft which fans out at the surface to create that stronger west wind. Because the most spectacular accidents have involved wind shear in or near thunderstorms, the nature in which a thunderstorm builds and operates is important to the subject.

Several factors contribute to the development of a thunderstorm. There has to be a mechanism to lift air. It can be the heat of a hot summer afternoon, it can be air flowing over mountains, it can be the lifting that is found ahead of an advancing low-pressure storm system, or it can be the lifting that is associated with a cold front pushing into an area, or warm air moved aloft over colder air north of a warm front.

Instability is another ingredient. Basically, the air is considered to be conditionally unstable if the temperature drops more than two degrees centigrade per 1,000 feet of altitude. Moisture is also required, which means a low-level flow of moist air, which usually moves over land from some large body of water. In the classic strong thunderstorms that romp around the U.S. middle-section in the springtime, a southerly flow from over the Gulf of Mexico provides the moisture.

When air is lifted it expands and cools. When it cools to the dewpoint (the temperature at which the air becomes saturated), condensation occurs and clouds form. If the air is stable, that is, if it cools slowly the higher you go, nothing much happens after the clouds form. If the air cools rapidly with altitude, then the clouds will continue to develop vertically.

Some heat is generated in the condensation process so the air in the clouds will cool less rapidly than the surrounding air. Because the moist air that is building in the clouds is warmer than the air around it, the tendency is for it to continue rising. It has become unstable in relation to its surroundings. The more moisture that is drawn into the cloud, the more heat is generated by condensation. It builds on itself: as the flow of moist air into the building clouds increases, it gives even more energy to the building clouds. As the clouds build, the moisture particles that are being drawn in and turned to cloud collide and become ever larger, and eventually the size of the drops and the amount of moisture in the clouds will become more than the upward motion can support. Rain starts to fall, bringing down with it cool air from high in the clouds. At the moment the rain starts (and the lightning begins, which is usually about the same time) and the air starts flowing down in the center of the storm, the inflow and upward flow might actually increase in velocity around the downdraft in the center The descending air is called a downdraft, downburst, or microburst.

## *Thunderstorms*

So here you have this humongous cloud that has built into a thunderstorm. Rain is pouring out of the middle, air is rushing downward and fanning out at the surface, but at the same time a large amount of air is being drawn into the storm. This air, around the edges, moves up and over the cooler air that is coming out of the storm. If you are standing at the surface, facing a storm to the west, there's likely a strong and cool breeze in your face. But a few hundred feet above where you stand the wind might well be from the opposite direction and be warm and humid.

That is a simplistic explanation of how a storm is built, but it should serve to outline the mechanism that results in strong winds blowing in different directions over either a short vertical or horizontal distance.

Thunderstorms can form in lines or clusters, and each thunderstorm cell has a finite life that is determined by the strength of all the factors and the length of time it takes for the moisture to be removed from the cloud by the rain process. One might last for 30 minutes or an hour, but if all the ingredients that built a storm remain, then continuous thunderstorm cell generation in the area might give every appearance of thunderstorm activity lasting for longer than the life of an individual cell.

## An Oshkosh Storm

I saw a perfect example of the severe wind shear that can accompany strong thunderstorms one summer evening, while attending the big air show at Oshkosh, Wisconsin.

My activity for the evening program was to narrate a slide show about aviation weather. The evening programs are held in large tents, seating a few hundred people, and as the time for the program approached, the thunderstorms associated with a stationary front just north of the airport appeared that they would help illustrate the talk on weather.

The conditions got downright mean. The storms moved south, and an extremely strong outflow developed. The wind blew out of the north with enough velocity to tip over some airplanes, and we decided to wait a minute before starting the program. The show outside was better and I was standing watching the weather when the wind shifted direction 180 degrees and the temperature jumped markedly. The storm, which was maturing, had apparently decided to take one more deep breath before really getting to work. That happened again. The wind shifted back to cool and northerly and then back to warm and southerly one more time before it settled down to blow out of the north and a torrential rain went to work on the countryside. A spectacular lightning show developed and a couple of times the tents seemed in jeopardy. It was quite a storm, and I always think about the way that wind behaved over perhaps a 15-minute period at the beginning of the event. The cloud was alive, and the amount of energy that it was generating was moving air in and expelling it with considerable force.

Given that any experienced pilot would have at least a rudimentary understanding of how a thunderstorm works, and that most have watched storms from the ground and know of their strength, the question is why some pilots flirt with the things when approaching or departing.

The root of this is probably found in the capricious nature of thunderstorms. Using the one at Oshkosh as an example, that storm was creating an unusually severe shear as it huffed and puffed. I have been an avid weather watcher for years and lived a good number of years in thunderstorm country, but I had never before seen one that behaved quite like that one. Where it did not have a particularly different appearance, the outflow and inflow behaved differently. Another phenomenon that I have seen only once is a strong thunderstorm that moves toward the west (in the Northern Hemisphere). This is unusual because thunderstorms move with and are steered by the middle-level winds (those about 18,000 feet above the surface) and the middle-level winds are not often out of the east when conditions are conducive to thunderstorm development. In either unusual case, had I been in an airplane instead of watching from on the ground, the experience (and the behavior of the storm) would have been something new to me. You might say the potential would have existed for a surprise, and surprises are the worst things in the world for an aviator.

## "Normal Ops"

When there is a thunderstorm-related accident around an airport, the continuation of "normal" operations can be another factor in prompting a pilot to continue an approach or initiate a departure. We'll examine this in relation to accidents that have been reported on by the National Transportation Safety Board. Also think back to the transcript of cockpit voice recorder in the Dallas/Fort Worth crash. The pilots of the Delta flight were following other aircraft. The other aircraft (one was a smaller airplane, a Learjet) were flying through whatever was there and were landing safely. Is that not enough to prompt a person to continue an approach?

Much of what we do in airplanes has a basis of personal experience as well as the experiences of other pilots. Over the years, a strong feeling built on experience has been that thunderstorms are more manageable when the airplane is flying low as opposed to high. Couple this with the fact that an airplane on approach or about to depart often has but two choices—either divert the flight (or delay departure) or follow the prescribed path—and some understanding of why pilots continue might be gained.

### THE OTHER SIDE

Airline pilots will adamantly contend that the pressure of making the schedule has nothing to do with a decision on weather in a terminal area. The sincerity with which that statement is often made is convincing, and I believe it. A pilot is, after all, always the first to reach the scene of the crash and is located in a part of the airplane that is as vulnerable as, if not more vulnerable than, any other. All the pilots I know fully realize that parking a large airplane anywhere other than at the gate involves a very high risk, and takers of big risks don't last long in flying. In the airplane business they are not likely to last long enough to get command of an airliner.

## *Large v. Small*

The nature of airplanes has something to do with all this, too. Generally, when considering weather, the large airplane has all the advantages over a smaller airplane. It is uncomfortable, but a transport category airplane can be flown through a thunderstorm at altitude and many of them have been flown through thunderstorms. In jet airline operations in the United States, I know of but one case of a thunderstorm encounter causing the structure of the airplane to fail—a British-built BAC 1-11 Braniff ship years ago. (There was also the case of hail and rain causing both engines to fail on the Southern Airways DC-9 at Atlanta, but the airframe made it through the storm okay.) I've ridden through a thunderstorm in the cockpit of a Boeing 727, and while the ride was sure bumpy and the crew was working quite hard, the issue never seemed in doubt. (The captain later wrote up the airborne weather radar on the aircraft as being unreliable.) But one place the big airplane loses its advantage over the small is in the low-altitude wind shear encounter. The mass and inertia of the large airplane, as well as the amount of aerodynamic drag that is associated with its flaps system (to allow flight at the slower speeds required for an approach and landing), means that the airplane has poorer acceleration qualities than a light airplane. In a jet, for example, to fly down a normal glideslope to a landing, the power required might be as high as 60 or 70 percent of that available. In a light airplane, it would be more like 30 percent. So in the jet the pilot has less to work with and has a far larger machine to try to make adapt to a new wind scenario.


### TWINS

If there is a best airline airplane to fly through a wind shear, it would be one with two engines. Because these airplanes must meet stringent one-engine-out climb requirements, they have an excess amount of power with both engines operating. (A three- or four-engine airplane has less excess power because with one engine out it loses only a third or a fourth of its available power, where the twin loses half.)

Before going into case histories, let's visualize how a typical thunderstorm might affect an airplane that is flying through the storm on an approach. Remember, what is happening to the airplane is a product of changing wind, or wind shear.

To begin, the airplane is flying toward the storm, which is between it and the airfield. The airspeed is stabilized at 130 knots, and the speed of the aircraft over the ground is 130 knots, the wind aloft in this initial area being calm.

As the airplane approaches the storm, the updraft might be encountered, causing an increase in airspeed if the altitude is held constant. If the pilot were to hold altitude and airspeed, this would require a reduction in power and a lowering of the nose of the aircraft. Flight into the outflow of the storm would result in the airplane flying into an increasing wind from ahead. To counter the initial effect of this, which would be an increase in airspeed, the pilot would have to reduce power a bit more and lower the nose of the aircraft.

An airplane's speed across the ground is equal to its true airspeed minus any headwind component or plus any tailwind component. We were flying at 130 knots; the increase in headwind from the outflow of the storm was 30 knots so the groundspeed of the airplane will be reduced to 100 knots.

### THE BAD NEWS

Nothing wrong so far, but something bad is about to happen to our airplane. After it moves into the updraft and the outflow, the next move has to be into the downdraft in the center of the cloud, usually associated with the rain, and then, as the airplane continues, the headwind of the outflow will turn into a tailwind from the outflow as the airplane moves to the other side of the storm.

In a downdraft, the airplane is flying in sinking air and in effect it has to climb to stay even. The pilot would have to increase power far in excess of what he had been using (just a moment ago power was reduced and the nose lowered to keep the airplane on speed and altitude or glideslope) and the pilot would have to bring the nose of the aircraft higher to generate more aerodynamic lift. That's trouble enough; the kicker is the change from a headwind to a tailwind. Where the groundspeed was 100 knots with a 30 knot headwind, the groundspeed would have to increase to 160 knots with a 30 knot tailwind. And when the aircraft goes rapidly from one wind situation to the other it must, in effect, accelerate the amount of the change, or 60 knots in this example. This takes a lot of power in addition to the power that was already added to counter the effect of the downdraft. The result can be a substantial loss of airspeed or altitude or both.

Some thunderstorms are weak and some are strong. A pilot might fly through one, seeing all those factors work on his airplane but in a fashion that is manageable given the power available. Then, the pilot might try a storm that has just reached maturity, with an outflow (and thus, an effect on the aircraft) that has become markedly stronger—so strong that the aircraft does not have the performance to cope with the wind shear. In the accidents that have happened, it's a cinch that sooner or later in the sequence the pilot increased the output of the engines to maximum available power, yet the airplane still settled or flew into the ground.

The airlines and business jet operators that use wind shear models in simulators train their pilots in recovery techniques to use if wind shear is encountered. I have flown most of the shear models in simulators ranging from the Learjet up through the Boeing 757, and if the proper procedures are followed it is possible to fly through a strong shear without hitting the ground. And there are documented cases of airline aircraft encountering strong wind shear on approach, and the crew managing to extricate the airplane by using the simulator-tested procedures. But as good as flight simulators are, there remains the fact that when you set out to practice wind shear flying in a simulator, you know what is coming. In the airplane, a pilot who pays attention knows what is coming as well, but he doesn't know exactly when it is coming or what strength it will be when encountered. Some airlines are stressing that crews should abandon the approach at the first sign of the effects of a thunderstorm on final approach. If the airplane starts to show the influence of an updraft and flight into the outflow, for example, as shown by an increase in airspeed and altitude, the drill is to get out then without waiting for the bad stuff, the decrease in airspeed and altitude, to begin. It's still a highly subjective call. Maybe it's okay to continue if there is a little bit, but not okay to continue if there is a lot. How much is a little and how much is a lot? Also, there's always that possibility that the storm is strengthening at the very moment, and that it will increase in intensity as it is being penetrated with far stronger effects on the other side of the storm than on the side initially encountered.

## *Hardware*

However it is done, the call on a thunderstorm in the approach area is a close one. The pilot does have tools to use in making the decision. Airborne radar and lightning detection equipment, ground-based radar, visual sightings, and low-level wind shear alert systems are all available for use; a more sophisticated radar system, called Doppler radar, is becoming available in both ground-based and airborne versions to help the pilot avoid unfortunate encounters with low-level wind shear. (An airborne wind shear alert system is also being put into use by airlines and corporate users, but the aircraft has to begin a wind shear encounter for this to work. It won't tell that there is wind shear ahead; it will warn that shear is being encountered and will do so more quickly than human senses.)

Airborne weather radar and lightning detection equipment are very effective in identifying a thunderstorm. With radar, which shows rainfall, the pattern of the return on the scope can be effective in identifying a thunderstorm, and the potential for wind shear. Simply put, if the picture is one of no rainfall and then, in a very short distance, heavy rainfall, there is a high likelihood of convection activity and shear. Most radar sets use three colors to depict rainfall rates: red is the heaviest rain and if a pilot sees red ahead on final approach, a wind shear pattern is likely to play on the airplane before landing.

Lightning detection equipment—the 3M Stormscope pioneered in this area—gives the azimuth and approximate distance of lightning activity. If there is lightning there's almost certainly some wind shear.

The plain old visual sighting by the pilot can count for a lot, too. On the Dallas/Fort Worth transcript, the first officer remarked that there was lightning ahead. The crew certainly didn't need any sophisticated electronic gear to *see* lightning ahead. But it goes far past that. There are other things to consider because you can often fly under a cloud that is producing lightning without encountering significant low-level wind shear. (This is true when the base of the thunderstorm is high, when there is a layer of colder air near the surface with warm air flowing above it, and when the outflow of the thunderstorm doesn't penetrate into the colder air at the surface.) So the appearance of the cloud which is producing the rain and the lightning has to be considered. Bad storms generate some low-level churning clouds that are clear evidence of a lot of action. However, a pilot can visually examine a thunderstorm from only one side at a time. Approaching one from the north, for example, the pilot might be looking at the more benign side of the storm. The other side, the side out which the airplane would fly, might have an ominous and mean appearance. Any pilot who has flown a lot in weather has looked back at a cloud that beat up and hosed down his airplane and thought that he sure wouldn't have flown through that one if he had had knowledge of what the other side looked like before poking into the beast.

Even with airborne weather radar, a pilot might not be able to make an accurate assessment of the other side of a storm ahead. Radar signals are attenuated, that is, used up, as they penetrate into an area of rain. The farther into a rain area you are looking, the less accurate the picture becomes. If a storm is a few miles thick, the radar information the pilot gets about the other side of the storm, or about another storm farther away but behind the first storm, might be highly inaccurate.

Doppler radar is available in both airborne and ground-based versions and has a substantial advantage over regular weather radar because it can "see" wind. Doppler measures the movement of particles toward or away from the radar site; this shows where the wind is shifting and where the wind shear might exist. A fine tool, Doppler isn't a total and final answer simply because of the quickness that characterizes the changes found in and around some storms. A storm might look okay on a Doppler radar when the airplane is five miles from the airport, but might change a moment later.

### Ground Equipment

Air traffic control radar can "see" precipitation but because heavy precip interferes with the controller's ability to see airplanes on the scope, circuitry is employed that minimizes the return of precipitation. The traffic radar is not designed to make weather interpretation easy, and controllers vary in their use of this equipment to advise pilots of weather. It's not their job, and if asked about weather, some controllers simply tell the pilot that their radar isn't adequate for giving weather information. Others develop the ability to interpret the returns to some extent and try to give pilots the benefit of what they see. But it must always be understood that the controller's job is to keep airplanes separated from one another, and to keep traffic moving. They have no responsibility, legal or moral, to separate airplanes from weather. That's the purview of the pilot.

There's a low-level wind shear alert system at major airports that is, in a sense, misnamed. These systems consist of anemometers (which give wind direction and velocity) at various locations on or near the airport, usually close to the final approach paths. If there is a substantial difference in wind at the various locations, this is taken as an indication of potential wind shear. The reason the system's name isn't quite right is that it is a surface system, and strictly a close-in system. The information is valuable to pilots as they actually fly the last half mile and land the airplane, but it provides no information about what's happening farther out on the approach or departure path. It's still better than no information, even though it should probably be called an airport wind *shift* warning system.

### Airborne Wind Shear Alert

Airborne wind shear alert systems are relatively new. They measure a lot of parameters and warn the pilot that the airplane is entering an area of changing wind. Control commands are given to help the pilot fly through the shear, and the system holds promise. But there can still be wind shears that exceed the performance capability of an airplane and the emphasis has to be more on pilots avoiding conditions where strong wind shear is a possibility. If a strong one is penetrated, the airplane and its occupants are at risk regardless of the equipment in the airplane or on the ground.

## Weasel Wording?

At this point, it might appear that this dissertation is dissolving into a collection of weasel words designed to take pilots off the hook. That is not the case. Even when the result is disaster, the pilot must feel in the beginning that it is prudent to proceed; the intention is to try to illustrate why the pilot feels this way.

## New Orleans

Wind shear can get airplanes on takeoff as well as on approach. The Pan American 727 accident at New Orleans is an example of a takeoff shear encounter.

Understand to begin with that the decision to take off, or to delay, is totally the pilot's. It is his job to look, evaluate, and make the decision. The airlines and corporate operators using crews (as opposed to one pilot in many propeller airplanes) urge first officers to be a part of the decision process; sometimes they are taken into the loop and sometimes they are not. That depends on the captain. The input from another crewmember sure can't hurt, but the captain's job is to make the final decision.

### THE WEATHER

The weather wasn't expected to be bad in New Orleans the day of the accident. The area was under the influence of high pressure and there were no fronts or low-pressure systems within 100 miles of the airport. There were no convective sigmets, which are weather advisories issued for strong thunderstorms or those imbedded in other clouds. There was a sigmet for thunderstorms to the east of New Orleans with little movement projected for this activity. However, the actual weather at New Orleans as the crew prepared to take off was a different matter. At the airport, the official observation reported a relatively high ceiling but with rainshowers and a cumulonimbus (thunderstorm) overhead.

Thirty minutes before the accident a National Weather Service radar site 30 miles northeast of the New Orleans airport reported that there was 30 percent coverage of thunderstorms with intense rainshowers over the area, with the highest thunderstorm top (a possible indication of severity) 40 miles northeast of the radar site. Observers around the airport reported and measured very large amounts of rain (over two inches) that afternoon, with one observer estimating that the majority of the 1.75 inches of rain measured at his location fell just before, during, and immediately after the crash.

### FLIGHTCREW OBSERVATIONS

The crew of another airline flight reported that there was a thunderstorm cell right over the airport when they departed, and another crew reported that the largest cell was east-northeast of the airport and that the gradient (the rate of change in rainfall with distance) in this cell was "very steep." That would indicate the possibility of a strong storm. The captain of this flight (which departed about seven minutes before the accident) reported that they encountered heavy rain and wind shear about halfway down Runway 19. This shear was reported to the departure controller. Another airliner that took off a few minutes later on the same runway, 19, reported no shear or turbulence.

There was a lot of discussion between pilots and controllers about which runway would be best to use, and on the Pan Am flight deck the choice of runways was important because of the temperature and the weight of the aircraft. As they were preparing for departure, the first officer remarked, "Any more than one knot of tailwind and we wouldn't be legal for 15." As a result of that, and because of the reported northeast wind, the crew informed the controller that they would need Runway 10 for departure. The captain briefed the first officer, who would be flying the airplane on the takeoff, on the procedures to use for a "heavy takeoff" and told the flight engineer that if they had an engine failure after the takeoff decision speed, that he should start dumping fuel to lighten the Boeing 727.

Following are some pertinent communications between the controllers and this airplane, between the controllers and other airplanes, and between the Pan Am crewmembers. Italicized portions indicate intracockpit conversation.

N1MT: And mike tango, what's that wind doing now, please?

GROUND CONTROL: Wind, ah, 060 degrees at 15, peak gusts 25, low-level wind shear alert at, at northeast quadrant, 330 degrees at 10, northwest quadrant 130 degrees at three.

SOUTHWEST 860: Ground, Southwest 860, with the present wind conditions we're requesting 28 for departure.

GROUND CONTROL: Southwest 860, roger, see what we can work for you.

N1MT: And, ah, ground 31 mike tango is also requesting 28.

PAN AM CAPTAIN: *Now we might have to turn around and come back.*

PAN AM FIRST OFFICER: *Yeah.*

PAN AM FIRST OFFICER: What are your winds now?

GROUND CONTROL: Winds now 070 at 17 and, ah, peak gust that was, ah, 23 and we have, ah, low-level wind shear alerts all quadrants, appears the frontal passing overhead right now, we're right in the middle of everything.

PAN AM CAPTAIN: *Let your airspeed build up on takeoff, takeoff.*
PAN AM CAPTAIN: *I don't understand why these guys are requesting Runway 28.*
PAN AM FIRST OFFICER: *I don't either. Must be sittin' there lookin at a windsock.*

The passengers were informed over the PA system that they would be departing shortly and that the aircraft would be circumnavigating some "little thundershowers out there." The aircraft was then cleared for takeoff. As they were finishing up the takeoff checklist and preparing to go, the tower reported to an arriving aircraft that a 10-knot wind shear had been encountered by an aircraft 100 feet high on final approach.

PAN AM FIRST OFFICER: *We're cleared for takeoff.*
PAN AM CAPTAIN: *I would . . .*
PAN AM FLIGHT ENGINEER: *Lookin' good.*
PAN AM CAPTAIN: *A slight turn over to the left.*
PAN AM FIRST OFFICER: *Takeoff thrust . . . wipers.*

The captain apparently called out the appropriate speeds on takeoff, though these couldn't be positively identified on the recording.

PAN AM CAPTAIN: *Positive climb.*
PAN AM FIRST OFFICER: *Gear up.*

The conversation on the remainder of the tape couldn't be positively interpreted but the captain apparently said to the first officer that the aircraft was sinking and that he should come on back, in reference to the control column. The ground proximity warning sounded, with its "whoop, whoop, pull up," and the last sound was that of the impact.

The airplane initially hit trees located almost a half-mile past the end of Runway 10. The airplane then struck a second group of trees 300 feet farther east and finally hit trees and houses before coming to rest almost 5,000 feet from the end of the runway. All aboard died.

The National Transportation Safety Board report on this accident was fascinating in that it broke with tradition and did not find fault with the pilot-in-command. The safety board's probable cause was "the airplane's encounter during the liftoff and initial climb phase of flight with a microburst-induced wind shear which imposed a downdraft and a decreasing headwind, the effects of which the pilot would have had difficulty recognizing and reacting to in time for the airplane's descent to be arrested before its impact with the trees. Contributing to the accident was the limited capability of current ground-based low-level wind shear detection technology to provide definitive guidance for controllers and pilots for use in avoiding low-level wind shear encounters." Among the conclusions was this: "The captain's decision to take off was reasonable in light of the information that was available to him."

### EFFECT OF HEAVY RAIN

A detailed performance analysis was done after this accident and the theory of the effect of heavy rain on an aircraft was explored. Basically the rain theory says that an airplane in heavy rain is affected in three ways. Some amount of rain adheres to the airplane, effectively increasing its weight. The raindrops hitting the airplane must assume the speed of the airplane and the exchange of momentum retards the velocity of the airplane. And the rain flowing back on the wings forms a water film, roughening the surface and reducing the aerodynamic efficiency of the wings. Theoretical studies have been done to determine the effect of rain on aircraft performance. According to testimony at the hearing on the New Orleans crash, NASA has reviewed the data in the rainfall study. Based on its review, NASA concluded that there is not (or was not at that time) "enough data to determine whether the estimates postulated therein were either reasonable or unreasonable." I think that any pilot who has made takeoffs in heavy rain would suggest that there probably is some effect.

A detailed analysis of the surface and low-level winds that might have existed at the time of the accident was conducted by the National Oceanic and Atmospheric Administration (NOAA). The analysis was based on an evaluation of large-scale meteorological patterns, satellite data, weather radar data, and witness accounts of the weather. The analysis concluded that the airplane flew through the center of a convectively generated downdraft shortly after it lifted off. According to this analysis, the airplane also flew through "an adverse wind shear of 39 knots." The maximum downflow was said to be seven feet per second at 100 feet. In a separate study, funded by Pan American, the location of the center of the downdraft (called a microburst) was different from that suggested by the study done by NOAA. But there was no question that the airplane started its takeoff roll with a headwind that changed to a tailwind after the aircraft lifted off, and that it also flew into a downdraft.

An airplane on takeoff is at a greater disadvantage in a wind shear than an airplane on approach for several reasons. The power is already at or near a maximum value. The airplane has little or no altitude. The airplane weighs more because it has fuel on board for the trip at hand. And the airplane is in a relatively high drag condition with a goal of both climbing and accelerating. Wind shear will affect both adversely.

At the request of the NTSB, Boeing analyzed the information from the flight data recorder. Because the recorder on this aircraft didn't record flight control inputs, engine thrust inputs, longitudinal acceleration, and airplane pitch angles, assumptions concerning these items were required.

Boeing divided the analysis into three parts: ground roll to rotation (the initiation of the lift-off); the takeoff, including rotation, lift-off, and climb to 35 feet; and the remainder of the flight.

Thirteen possible scenarios were analyzed and the finding was "that the horizontal wind changed from a headwind or slight tailwind when the airplane was 35 feet high to an increasing tailwind which then diminished slightly before initial contact with the trees. The vertical wind increased from a slight downdraft at 35 feet AGL (above ground level) to a maximum downdraft as the aircraft reached 100 feet AGL." In the Pan Am analysis, the computations showed that the airplane reached a maximum altitude of 163.2 feet and encountered a maximum headwind of 17 knots and a maximum tailwind of 31 knots.

### CONTROLLERS

The controllers in the tower stated or testified that it was raining on the airport when the airplane departed. The weather was being depicted on a radar scope in the tower but the senior controller in charge said that "it didn't appear significant enough to affect aircraft operations." All the controllers who were on duty said that the weather at the time of the accident was typical of thunderstorm weather which occurred during a summer day at the airport and National Weather Service data showed that during the preceding 17 years there was an average of 13.47 days in July with thunderstorms reported at the airport.

The captain, who had worked for Pan Am or National Airlines (which was bought by Pan Am) for 17 years, and the first officer, who had worked for the airline for over five years, had both been based in Miami since being hired. Evidence in the investigation showed that, for the most part, they had flown routes which traversed the southern tier of the U.S., where thunderstorm weather is common during the summer. According to the report, "Thus the evidence was conclusive that both the captain and the first officer were familiar with the air mass type thunderstorm weather that was affecting the New Orleans area and airport on the day of the accident. The evidence also indicated that they most probably had landed and departed from airports under weather conditions similar to that which existed at New Orleans International Airport on July 9, 1982." The report further stated: "There is no evidence that management exerted any pressure on its flightcrew to keep to schedules in disregard of weather or other safety considerations."

### SPLIT SECOND

To illustrate what a split-second business this is, the following is from the conclusions in this accident report: "The performance analysis indicated that, at 5.9 seconds before initial impact, had the pilot been able to increase the airplane's pitch attitude and maintain the indicated airspeed that existed at that time, Clipper 759 (the aircraft's radio call sign) theoretically would have been able to maintain an altitude of 95 feet AGL."

That is an interesting thing to contemplate because it shows the depth of the investigation into an accident like this, as well as the overpowering influence of the weather condition. When it gets down to a suggestion that, had the pilot done something absolutely correctly at a point identified by tenths of seconds, he would have cleared 50-foot-high trees by 45 feet, it is reflective both of a detailed analysis and a suggestion that the window of success wasn't even a peephole. Given the conditions that existed, the chances of this airplane making it were slim indeed.

## This One Made It

Another 727, a United airplane, struck an antenna at the far end of the runway at Denver while taking off. The airplane was successfully landed after the strike and, in this accident, the NTSB found the probable cause to be "an encounter with severe wind shear from microburst activity following the captain's decision to take off under meteorological conditions conducive to severe wind shear." Factors that were listed included the limitations of the low-level wind shear system, the terminology used by the controller in relaying information, the captain's erroneous assessment of a wind shear report from a turboprop airplane, the fact that the captain didn't get a wind shear report from an airplane similar to his because of congestion on the air traffic control frequency, successful takeoffs made by other airplanes, and the captain's previous experience operating at the airport under wind shear conditions.

Denver can have weather that is unusual in the sense that it is convective, it does result in downdrafts, but the activity doesn't look like the thunderstorms that most of us are used to seeing in flat country. The bases of these storms are high, and the moisture content is lower than in storms elsewhere. They form over the front range of the Rockies, move out over the Denver area, and usually dissipate shortly afterwards. Often the rainfall associated with this activity is reported as "virga"—that is, rain that evaporates in the hot and dry air beneath the storm before it reaches the ground. Nevertheless, when the cumulonimbus matures, and the moisture begins to fall, the downdraft, or microburst, can both reach the surface and be quite strong.

There was a high overcast with some virga in the area as this Boeing 727 waited its turn for takeoff. One preceding airplane, a four-engine turboprop, gave a report of a 25-knot airspeed loss on takeoff at about 200 feet and the controller gave this report to a light twin, a business jet, and an airline 737 that followed in sequence. Where the surface wind had been light earlier, in the five minutes before the departure of the aircraft involved in the accident, the controller started reporting westerly winds as strong as 33 knots. (The aircraft were taking off to the north.) The controller asked the departing Boeing 737 crew if they had encountered wind shear and they replied "negative." Other flights were being cleared for takeoff all the while.

The United captain involved said that he heard the wind shear report from the turboprop, and had made a mental note to listen for a report from the MD-80 that preceded his flight. The captain recalled saying to his first and second officers that the "Frontier (the MD-80) that preceded him on takeoff didn't say anything, but I think in light of the other report, even though it was a smaller aircraft and he was airborne, we'll climb out at V2 plus 20." (V2 is the speed that would be used in climb immediately after takeoff; adding to that would increase the safety margin in case of a wind shear encounter.) The captain said that he didn't ask for a report from the MD-80 because of congestion on the frequency.

After the flight had been cleared to taxi into position on the departure runway to hold for takeoff clearance, the flightcrew said they observed dust blowing from west to east across the runway near midfield. Another flight, preparing to take off on the other runway parallel to this one, reported, "Our sock sitting in front of us gives us a pretty good tailwind so we are not ready to go yet." ("Sock" was in reference to a wind sock near where they were waiting to take off.)

The controller's clearance to the accident aircraft went like this: "United 663 center field wind 280 at 22, gust to 34, north boundary wind 280 at niner, numerous wind shears in three different quadrants, 35 left, cleared for takeoff." The crew said that they didn't recall hearing the controller's mention of wind shears or the transmission from the flight on the other runway.

The second officer (flight engineer) reported that after they started rolling he heard an airplane that departed ahead of them report a hesitation in acceleration on the takeoff roll at midfield, "or something to that effect." This report had actually been made by the crew of another 727 which had departed on the same runway about a minute earlier. About halfway down the runway the first officer reported to the captain, who was handling the controls, that they were slow in accelerating. Then, according to his recollection, the airplane began accelerating normally. The captain said he associated this with the midfield crosswind observed earlier and that he momentarily considered aborting the takeoff when the airspeed began to hesitate, but disregarded the thought when the airspeed began increasing.

As the captain began rotating the aircraft for takeoff at a speed of about 141 knots, the first officer reported, "Your airspeed is falling off." According to the captain, this was followed immediately by, "You've lost 20 knots," from the first officer. The captain checked the rotation to a deck angle lower than normal to reduce aerodynamic drag and enhance acceleration, and applied full power to the engine to regain the proper speed for rotation.

The captain said that as the speed started increasing toward the proper value for rotation and lift-off, the marks indicating that there were 2,000 feet of runway remaining flashed by the corner of his eye. He resumed rotating the airplane to the proper attitude for climb. The airplane lifted off, and in the initial climb, experienced a 30-knot fluctuation in airspeed. As the airplane was climbing through 8,500 feet, the crew found that they couldn't pressurize the cabin so they returned and landed. Though they thought they had cleared the localizer antenna off the end of the runway, they had actually hit it, leaving a four-inch-by-five-foot gash in the aft fuselage. That was why the cabin wouldn't pressurize. The only damage other than to the antenna and the fuselage was to the grass. It was scorched (by the jet exhaust) from a point 300 feet past the end of the runway to a point 245 feet in front of the platform on which the antenna was mounted. That's close.

Some of the information from the report on this accident gives a clear picture about how quickly wind conditions can change in the area of a downdraft.

The airplane was taking off toward the north, runway bearing of 350 degrees. Wind from a direction more southerly than 260 degrees would constitute a tailwind component. The centerfield wind given the crew before takeoff was from 280 degrees, strong and gusty. The north boundary wind was from the same direction but relatively light. However, as the crew was conducting the takeoff, the *northeast* boundary wind detector went from 220 degrees at 15 knots up to 220 degrees at 40 knots. It was down to 23 knots a couple of minutes later.

Of all the aircraft departing in the immediate time before the accident, none reported any serious trouble but all did encounter some wind shear effect. One first officer remarked, "Airspeed losses are common at Denver." The airplane involved in the accident was loaded to about 95 percent of its maximum allowable weight for takeoff, given the temperature and runway length available. Some of the other aircraft were operating at a smaller percentage of their maximum allowable weight for the existing conditions, so they would have been better able to cope with wind shear.

In an analysis of the airplane's takeoff performance, the NTSB determined that the wind component along the runway sheared from an eight-knot headwind to about a 56-knot tailwind over a 44-second period during the takeoff roll, which took 62 seconds. (If you, like many others, time takeoffs on airliners, you'd have been bracing yourself by 62 seconds.) This is significant. It means that to maintain its acceleration in terms of airspeed, what an airplane has to have to fly, the 727 would have had to accelerate an extra 64 knots during this time, that being the product of the lost headwind and the acquired tailwind. The airport at Denver is about a mile above sea level, which affects aircraft performance, and the aircraft was flying at a relatively high weight. It was a difficult problem for the crew to work and for the airplane to fly.

## 727 On Approach

The fatal crash of a 727 at New York's Kennedy airport in 1975 prompted the most detailed analysis of thunderstorm-related wind shear that I had seen up to that time. It also brought on some new terminology—the one that stuck was *downburst*, later expanded to *microburst*. The effect of this phenomenon on the airplane involved in the accident, as well as those preceding it, was carefully documented, clearly showing that one man's bumpy passage might prove to be another's mission impossible.

Eastern 66 was one of a number of aircraft that were being vectored for an ILS approach to Runway 22L. The Kennedy final controller (the last to handle aircraft before turning control over to the tower for landing clearance) was vectoring the flight around thunderstorms and fitting it into the sequence with other traffic. The crew discussed the problems associated with carrying minimum fuel loads when confronted with delays in terminal areas, and one said he was going to check the weather at an alternate airport.

Ahead, an Eastern L-1011 had abandoned its approach to Runway 22L and reported, "We had . . . a pretty good shear pulling us to the right and . . . down and visibility was nil, nil out over the marker. . . . Correction . . . at 200 feet it was . . . nothing." This report was recorded on 66's cockpit voice recorder and the captain of 66 said, "You know, this is asinine." An unidentified crewmember responded, "I wonder if they are covering for themselves."

The controller asked the crew of 66 if they had heard the wind shear report and they replied that they had. He then established the flight's position as being five miles from the outer marker (about 10 miles from the airport) and cleared 66 for an approach. The first officer, who was flying the aircraft, called for the final checklist, and while it was being completed the captain reported that the airborne weather radar was on standby.

The final controller asked the crew of the Lockheed that had aborted its approach if they would "classify that as severe wind shift, correction, shear?" The flight responded, "Affirmative."

The first officer of 66 said, "Gonna keep a pretty healthy margin on this one." An unidentified crewmember said, "I . . . would suggest that you do." The first officer responded, "In case he's right."

After 66 passed the outer marker, about five miles from the airport, it was cleared to contact the control tower. The flight did and was cleared to land. On being asked about the ride on final by a National Airlines flight, the tower responded, "Eastern 66 and National 1004, the only adverse reports we've had about the approach is a wind shear on short final." National acknowledged this, Eastern did not.

On the flight deck, the following ensued as the aircraft proceeded toward the runway:

CAPTAIN: Stay on the gauges.
FIRST OFFICER: Oh yes, I'm right with it.
FLIGHT ENGINEER: Three greens, thirty degrees, final checklist.
CAPTAIN: I have approach lights.
FIRST OFFICER: Okay.
CAPTAIN: Stay on the gauges.
FIRST OFFICER: I'm with it.
CAPTAIN: Runway in sight.
FIRST OFFICER: I got it.
CAPTAIN: Got it?
FIRST OFFICER: Takeoff thrust.

There was an unintelligible exclamation simultaneous with the call for takeoff thrust, followed by the sound of the impact.

When the captain said that he had the approach lights in sight and then told the first officer to stay on the gauges, the aircraft was passing through 400 feet and the rate of descent was in the process of starting an increase from 675 feet per minute to 1,500 feet per minute. The aircraft began to move below the glideslope and then the airspeed decreased from 130 to 123 knots in two and a half seconds. The aircraft continued to deviate below the glideslope and was at 150 feet when the captain said he had the runway in sight. Four seconds then elapsed before the first officer called for takeoff thrust.

Witnesses near the end of the runway saw the aircraft flying in heavy rain at a low altitude. It first hit an approach light tower. Other witnesses reported the weather conditions when 66 passed over as follows: "Heavy rain was falling and there was lightning and thunder; the wind was blowing hard from directions ranging from north through east." (The airplane was flying toward the southwest on its approach.)

### NO, YES, YES, NO

Between the time that 902 aborted its approach and 66 crashed, two airplanes made successful approaches to Runway 22L at Kennedy. One of these aircraft was a DC-8; it flew through a wind shear about two miles out on final and lost 25 knots of airspeed but was able to continue and make a normal landing. A twin-engine Beech Baron was next and it flew through rain and light turbulence on a normal approach flown to the point where the airplane was 200 or 300 feet above the ground. There it encountered a high sink rate. The pilot applied power to recover from the sink and complete the approach without incident. The airport was closed after the Eastern crash so there were no other reports. But Dr. T. Theodore Fujita, a renowned meteorologist, obtained pilot reports from the eight aircraft that preceded 902, the Lockheed that aborted the approach. Most reported some wind shear but no great difficulty until the airplane immediately preceding 902. They had a rough time with it. The pilot reported that the aircraft encountered strong, sustained downflow from about 700 feet down to 200 feet. An abnormal amount of power was used for a long time, and below 200 feet a strong crosswind from the right was encountered, but there was practically no wind on the ground. (Dr. Fujita, professor of meteorology at the University of Chicago, has prepared a full and technically wonderful book entitled *The Downburst*. It is available from him at 5734 Ellis Avenue, Chicago, Illinois 60637 for $11.)

In its findings, the National Transportation Safety Board found that 66 encountered an increasing headwind of about 15 knots when penetrating the thunderstorm between 600 and 500 feet. At 500 feet it encountered a downdraft of about 16 feet per second. Then the headwind diminished about five knots and at 400 feet the downdraft increased to about 21 feet per second and the headwind decreased about 15 knots within four seconds. For the airplane to have remained on the glidescope at the selected airspeed, it would have had to accelerate about 20 knots while reducing its effective rate of descent by 21 feet per second (1,260 feet per minute). It didn't.

The NTSB determined "that the probable cause of this accident was the aircraft's encounter with adverse winds associated with a strong thunderstorm located astride the ILS localizer course, which resulted in a high descent rate into the nonfrangible approach light towers. The flightcrew's delayed recognition and correction of the high descent rate were probably associated with their reliance upon visual cues rather than on flight instrument references. However, the adverse winds might have been too severe for a successful approach and landing even had they relied upon and responded rapidly to the indications of the flight instruments."

The NTSB went on to state that a contribution to the accident was made by the continued use of Runway 22L when it should have become evident to both air traffic control personnel and the flightcrew that a severe weather hazard existed along the approach path.

## The Big Three

The accident report on the Delta Lockheed at Dallas/Fort Worth has not been issued at this writing, but based on the cockpit voice recorder information, and based on what happened, most observers feel that there is a strong similarity between that event and Eastern 66, with the exception that there were no reports of strong wind shear preceding Delta's approach. But the crew of both airplanes apparently knew there was convective activity— the Delta first officer mentioned seeing lightning ahead—and both continued the approach. For some reason the crew of 66 felt that perhaps the crew of the airplane ahead that aborted had erred. Perhaps the fact that two aircraft, one a light aircraft, subsequently landed strengthened this feeling.

When looking at the three notable wind shear accidents— notable because of the high loss of life—there are similarities. All were three-engine airplanes, and while a twin might have a better shot in a wind shear because of a potentially greater rate of climb, there have been plenty of twins involved in less spectacular wind shear accidents.

In each of the three, the first officer was flying the airplane. This has been much cussed and discussed in the aviation community and maybe there is something here. But this is true in other types of accidents as well; a discussion of it will be reserved for the chapter on coincidences.

A common factor that is interesting relates to the forecasts that were valid for the areas where the aircraft crashed at the times the captains were preparing to fly. When 66 left New Orleans, there was nothing in the forecast that indicated thunderstorms in the New York area on arrival. (The forecast was later amended to warn of thunderstorms but there is no evidence that the crew of 66 received the amendments that added thunderstorms.) In the New Orleans accident there was no convective sigmet warning of strong storms and it seemed a consensus that the day looked like a typical summer thundershower day. At Dallas/Fort Worth there was no strong frontal or organized storm activity.

In all three cases the airplanes flew into conditions that were in excess of, or at least nearly in excess of, the airplane's available performance. In each, the crew was working the problem. Perhaps the crew of 66 got onto it a little later than the others, but they were the first of the three to come a cropper, and, in the report on 66, the NTSB noted that little progress had been made in the formal training of flightcrews in the recognition of wind shear and the techniques for coping with wind shear. Training methods were developed after that accident, and flightcrews are now trained in the best way to cope with wind shear.

In examining the Eastern 66 accident, the NTSB explored the effect of air traffic congestion on the decision-making process. From the report: "When operating at capacity, the air traffic system in a high-density terminal area tends to resist changes that disrupt or further delay the orderly flow of traffic. Delays have a compounding effect unless they can be absorbed at departure terminals or within the en route system. Consequently, controllers and pilots tend to keep the traffic moving, particularly the arrival traffic because delays involve the consumption of fuel and tardy or missed connections with other flights, which would lead to further complications. As weather conditions worsen, the system becomes even less flexible."

There is truth to all that and the FAA has made an effort to address this problem. Now, if a major terminal is impacted by weather, they have that system of holding aircraft on the ground, out of the area. But it probably wouldn't have held 66, because no thunderstorms were forecast.

## Basics

So it's back to basics. In each case, the captain of the aircraft made the decision to continue. In each case, the decision was made based on broad experience. All the captains were veterans of the line. In no case did any other crewmember express apprehension about flying into the conditions. Nobody suggested that they go to an alternate, or, in the case of the one at New Orleans, not attempt the takeoff. The decision to continue, or to go, was simply not challenged. In each case the decision proved to be bad, but think how many good decisions were made in the 10 years of airline operation spanned by those accidents. If, as some suggest, pilots avoided operating within five miles of a thunderstorm cell, millions upon millions of passengers would have been inconvenienced in a 10-year period. To the families of the folks on the lost airplanes, it would have been well worth the inconvenience. For the rest it would have been a matter of throwing the air transportation system into chaos on a number of days during each year.

One final coincidence, mentioned only because it exists: New Orleans was involved in two of the three accidents—66 left there and Pan Am crashed there.

## What Next?

Based on what has happened, will all the new equipment and the training of flightcrews in wind shear flying techniques help? Maybe.

Doppler radar can show areas of wind shear, but these can develop and abate. I described the thunderstorm at Oshkosh that appeared to huff and puff, and from Dr. Fujita's research on the conditions at Kennedy the day that 66 crashed we saw a varying effect on the final approach course to Runway 22L. What might look okay to start might go bad when there is no time left to do anything but fly on through. Airborne wind shear warning systems might help by getting the pilot with the program more quickly than if he relies on his senses to detect the beginning of a wind shear encounter, and it might lead him to fly the precisely correct wind shear technique. But remember that split-second business in the New Orleans crash. Also, there's always the possibility that a pilot might feel he has an electronic crutch and depend on it to help him fly through something he might otherwise avoid. Used properly and cautiously, electronic warnings and information are good things but the real unanswered question is still about three well-intentioned but incorrect decisions in 10 years.

## Training

The wind shear training that is being done in flight simulators is, I think, valuable, though it isn't without a bit of controversy.

Some question whether the wind shear models that pilots fly through are accurate in relation to the shears that have caused accidents. Others question whether teaching pilots how to handle wind shear will result in more well-intentioned but incorrect decisions to continue into weather conditions that might exceed the performance capability of the airplane even if the wind shear technique is perfectly flown. Those are good questions, but the fact remains that flight simulators should be used to practice everything. Because wind shear is out there, the best technique of getting an airplane through one once encountered should be practiced.

One procedure or technique or policy that is finding some favor is one related to the sequence of events that might lead to the beginning of a wind shear. If the policy of "all that goes up must come down" were stressed, and approaches were abandoned at the first sign of shear, some advantage might accrue. Perhaps the first airspeed increase and high trend on the glideslope is not cause for an immediate go-around, but if an approach becomes markedly destabilized, the approach should be abandoned. A stabilized approach is basically one where the power setting, rate of descent, and airspeed are all within parameters, are stable, and the relationships are normal. In other words, you aren't going very fast and sinking slowly with the power set abnormally low. Or you aren't flying too slow and descending rapidly with an abnormally high power setting. Or you aren't transitioning from the first scenario to the last.

Would that someone comes up with a final answer to the wind shear question. The captain in the left front seat, faced with the lonely job of making the decision, would sure appreciate having a reliable solution. But all he's likely to get is better information on what might be ahead and better warning that the aircraft is starting to react to a wind shear, and better training on what to do next. Whether or not this will help has to be left to time for a grade.

## Other Wind Shear

One form of wind shear not related to thunderstorms has caused problems and accidents for years. This type of wind shear is related to a change in wind direction or speed with altitude and does not involve downdrafts, downbursts, or microbursts. It is thus more manageable, but that doesn't mean that it does not have to be managed. This type of shear has been responsible for airplanes touching the ground before reaching the runway, as well as for overrunning the runway on landing. Again, the larger and heavier the airplane, the more challenging the wind shear.

### EXAMPLE

The best example of this comes on a damp, dark, drizzly or rainy winter day, temperature relatively warm. In this condition, at most places, the wind at the surface will be light or from an easterly or southeasterly direction. However, aloft, a few thousand feet above the surface, the wind might well be rippling along out of the southwest at 30 or 40 knots, or, in extreme cases, even stronger. The strongest I have ever seen is a 60-knot difference between the surface wind and that at two or three thousand feet.

This has exactly the same effect on an airplane as the changing headwind or tailwind around a thunderstorm with two exceptions. One, as mentioned, is no downdraft. And, two, it has to be dealt with only once. That is, it won't be first an increasing headwind and then an increasing tailwind. It'll be a decreasing or increasing headwind or tailwind. It's relatively easy to know this is coming because a pilot knows his headwind or tailwind component at altitude and knows the surface wind. The difference between the two must be compensated for in the descent. For example, if landing to the northeast, surface wind calm, wind at 2,000 feet out of the southwest at 30, it'll be a decreasing tailwind of 30 knots in the descent. The airplane will, in effect, have to decelerate 30 knots while descending the last 2,000 feet and while flying the glideslope and maintaining a constant airspeed. This simply means that less power will be required to fly the glideslope and the proper airspeed. That sounds easy, but somewhere down around 50 feet the airplane will be through with the deceleration, it'll be down to the level where the wind is relatively calm, and if the power is not brought back up to the value for an approach without a decreasing tailwind the airplane will go low on the glideslope and might wind up in the approach light towers. It's the pilot's job to figure all that out in advance and to be ready.

## High-Altitude Wind Shear

High-altitude wind shear has not been a factor in large transport airplanes crashing, but it has been the cause of injuries to passengers and flight attendants and to stuff being thrown around the cabins of aircraft. Back in the days of prop airliners there was always talk of flying over the weather and turbulence, but as higher and higher altitude operations became commonplace with jets, it was found that the air isn't always smooth at any altitude, even in clear air.

The primary cause of turbulence in clear air is related to a change in wind velocity with altitude. On the TV they always show the jet stream. That in itself might not make turbulence because as it is shown it is more a reflection of the general high-altitude wind flow. But, especially in winter and sometimes in the spring and fall, there can be what's called a jet core, an area where the wind velocity is very much faster than the general wind in what's called the jet stream.

Say, for the sake of example, that the center of the jet core is at 29,000 feet and the wind there is out of the southwest at 150 knots. Down at 18,000 feet the wind is also out of the southwest, but at only 80 knots. That is a change of 70 knots in 11,000 feet. The effect of the higher velocity wind interacting with the lower velocity wind makes what might be called, for lack of a better term, burbles in the air and these are felt as turbulence. In rare cases, the velocity change can be abrupt enough for the airplane to react violently when flying into or within it. On most flights, the crew can adjust the altitude or the route to get a reasonably smooth ride.

One of the clearest examples of this I've seen was found when flying my pressurized light airplane in an area where there was a strong storm developing at the surface, beneath a jet core aloft—which is a common scenario when a strong storm is developing. My airplane will operate up to 23,000 feet, which would be below the center of most jet cores, but this day, on a northeastbound flight, I was enjoying a good portion of the tailwind. The ground-speed reflected a 100-knot tailwind, which was easy to live with. But from ahead came a report of severe turbulence at altitudes from 15,000 feet up to 29,000 feet. This turbulence was related to the interaction between the developing low-pressure system at the surface and the jet core aloft, and was apparently quite something. Before reaching the area, I descended to 13,000 feet, where the bumps were not bad but where the wind was from the south-east instead of southwest and my groundspeed was almost 100 knots less than it had been higher. The descent was far from smooth until the airplane was established in the area where the flow was lighter and from a different direction.

## *Downwind of Thunderstorms*

Another place the high altitude going can get rough is downwind of a thunderstorm when there is a very strong wind aloft. The vertical development of the thunderstorm disturbs the strong wind, much as a mountain disturbs wind and causes turbulence downwind of it. There have been cases of airliners encountering severe turbulence caused by this phenomenon. Most pilots give storms a wide berth when flying at altitude but apparently there are cases where, to get a smooth ride, you would have to go around the storm on the other, upwind, side.

Seldom is a flight flown when the pilot doesn't have to think about wind. It affects the amount of time (and thus fuel) that will be required to fly a flight and it is a factor in takeoffs and landings, climbs and descents, and comfort in all phases of flight. It would be nicer if we didn't have to deal with wind shear, but we do, and while there is no mystery there is plenty of challenge.

# 6

## Approaches

THE NEWSPAPER ACCOUNTS usually refer to a pilot (or an airplane) groping in fog to find a runway. How an airplane gropes can best be left to conjecture; the fact is that a lot of airplane accidents occur when the weather is bad. Often as not the pilot is indeed trying to land and frequently the ceiling and visibility at the airport are lower than that required by the instrument approach procedures. Or there might not even be an instrument approach procedure for the airport. In a television report of one crash, an airport employee said that the airport was covered with fog; the pilot made attempts to land, and finally requested that he place a car at the end of the runway and flash the lights. The pilot succeeded only in flying the airplane into the ground away from the runway.

Crashing while attempting a landing in low weather happens much more often to general aviation aircraft than to airline aircraft. There are several reasons for this. General aviation airplanes are more likely to be headed for an airport where there is no precision approach system. Lacking guidance in the vertical sense, a pilot can be tempted to sneak just a little bit lower in an attempt to gain sight of the ground. This has proven to be a fatal sneak. Where a general aviation airplane is flown by one pilot, which is usually the case, there isn't another person along to monitor and, unless the lone pilot is dedicated to abiding by the published and safe minimums, the tendency to fudge and go lower can be more pronounced. Finally, the general aviation pilot is usually flying the airplane for the purpose of getting to the destination at the appointed time. There's a business deal or personal reason for wanting to "be there," and if a pilot is thinking more about the press of completing the approach than he is concerned with following all procedures and abiding by all minimums and rules, then the potential for flying into the ground is great. The airline pilot, on the other hand, has at his side another crewmember and is being paid to fly the airplane. If it becomes necessary, because of weather, to divert, so be it. He gets paid for it. Also, by regulation the airline pilot isn't to begin an approach unless the reported weather is at or above the published landing minimums. A general aviation pilot is not bound by this rule. Weather reporting is not a prerequisite for an instrument approach in not-for-hire general aviation flying. That adds no necessary hazard to the operation—as long as the minimums are adhered to, an approach without weather reporting is just as safe as one with weather reporting. It's just that the temptation to fudge is both greater and punishable only by the airplane hitting something other than the runway.

Another tendency that claims a lot of general aviation airplanes every year is that second or third attempt after an approach is missed because of weather. If the pilot can't see the runway after flying down to the minimum altitude, the decision height in the case of a precision approach with glideslope guidance, or the minimum descent altitude in the case of an approach without glidepath guidance, and if the weather doesn't change, there's just no reason to think that the pilot will be able to see the runway on a subsequent approach—unless the pilot elects to descend below the minimum safe altitude. That is apparently what happens in a lot of these accidents. It is illegal but is generally known only to the pilot's conscience.

## Big Airplanes

There are cases of large airplanes crashing before reaching the runway but they are few, far between, and more likely related to wind shear than to the crew descending below a safe altitude accidentally or on purpose. You have to go back 10 or 15 years in the accident reports to find classic examples of this type tragedy.

As a chartered DC-9 was approaching Huntington, West Virginia, the weather report told of scattered clouds 300 feet above the ground, with broken clouds at 500 feet and an overcast at 1,100 feet. It was dark. The minimum descent altitude for the nonprecision approach to Runway 11 was 1,240 feet, 412 feet above the runway. The aircraft flew into trees on a hillside one mile west of the airport at an elevation of 912 feet. Why?

The National Transportation Safety Board's probable cause was "a descent below minimum descent altitude during a nonprecision approach under adverse operating conditions without visual contact with the runway environment." The safety board was unable to determine the reason for the descent though it found that the most likely explanation to be either improper use of cockpit instrumentation data, or an altimetry system error.

The cockpit voice recorder reflected a normal operation right up until impact.

FIRST OFFICER: A thousand feet above the ground, rate and speed good.
FIRST OFFICER: Speed a little fast, looks good . . . got bug and 12

[meaning that the speed was 12 knots above the calculated reference speed for the approach—not abnormal].

CAPTAIN: See something?

FIRST OFFICER: No, not yet. It's beginning to lighten up a little bit on the ground here at, ay, ah . . . 700 feet.

FIRST OFFICER: Bug and five.

FIRST OFFICER: We're 200 above.

THIRD CREWMEMBER: Bet'll be a missed approach.

FIRST OFFICER: Four hundred.

CAPTAIN: That the approach?

FIRST OFFICER: Yeah.

FIRST OFFICER: Hundred and twenty-six.

FIRST OFFICER: Hundred!


The aircraft hit the trees on the hillside less than a second later, and about seven seconds after the captain asked, "That the approach?" The NTSB construed the first officer's call of 400 to be related to height above the ground, perhaps as read on a radio altimeter that gives height above the ground. (The pressure altimeters read height above sea level and, at the minimum descent altitude, should have read 1,240 feet.)

The aircraft actually descended through the minimum safe altitude when it was two miles from the airport and continued descending until impact. Conversation between the captain and first officer just before the aircraft passed through the minimum descent altitude indicated that they were beginning to see lights on the ground, or at least the glow of lights in the clouds. The descent continued and ground witnesses saw the airplane flying clear of clouds two miles from the airport. Another factor could have been the difference in elevation of the area two miles from the airport and of the runway. There was a brightly-lit refinery there, at an elevation 300 feet lower than the airport, and if the aircraft had been flown down to 400 feet on the radio altimeter over the refinery it would have been at about the altitude of impact on the hillside. Still, the cockpit conversation did not reflect any awareness on the part of the crew that the airplane was being operated below the minimum descent altitude and, in fact, the investigation revealed that they had started the missed approach, per proper procedures, when the airplane hit the hillside. In its report, the safety board examined carefully the possibility of altimeter error and found that if the possible error examined did indeed occur, "the altimeter would have read 200 to 300 feet high, which in turn would account for the fact that the crew did not arrest the descent until the aircraft reached an altitude of approximately 916 feet m.s.l. (mean sea level) or over 300 feet below the MDA." However, the NTSB pointed out, such an error in altimetery would have been accompanied by an error in airspeed because they both operate off the same system. And there was no indication of an airspeed error.

### RADIO ALTITUDE

The NTSB also checked the correlation between the first officer's altitude callouts on the approach and what the radio altitude (above the ground) would have been at the time of the callout. Based on available information, they were close. Moreover, the NTSB reported, "The final exclamation recorded prior to the commencement of the sound of impact (hundred) accords with the altitude, which would have been reflected by the radio altimeter at that time and therefore is further evidence that the first officer may have been using that instrument during the approach." Still, the board found many flaws in the theory that the crew was using radio altitude instead of indicated altitude on the approach.

## Ground Proximity Warning

This accident and the crash of TWA 514 near Washington (see chapter 3) lent emphasis to the requirement for ground proximity warning systems in all air carrier aircraft, though when flying over lower terrain toward higher terrain, the system would give little warning since it is based on radio altitude above the ground. More important is the fact that most runways used by major carriers now have precision approaches, with glidepath guidance.

## Circling Approach

In that accident, the crew was making a nonprecision but straight-in approach to the runway. Even more challenging is a circling approach, where the aircraft is not aligned with the landing runway when it reaches the minimum descent altitude and must circle for alignment, beneath the clouds, while keeping the airport in view. This becomes even more challenging at night and is a procedure that most pilots don't like and that is shunned whenever possible by airlines operating large jets.

## Dark Night

A commuter airline de Havilland Twin Otter aircraft was making a nonprecision approach to Cape May, New Jersey, where the weather report was of a 400-foot ceiling and one mile visibility.

The minimum descent altitude for the approach was 440 feet (423 feet above the ground) for a straight-in approach and 480 feet (458 feet above the ground) for a circling approach with one mile visibility required in both cases. The approach procedure called for the aircraft to pass over the VOR navigational station 6.8 miles northeast of the airport and then descend to the MDA.

The airline's station manager radioed the weather report to the flight and said that he also told them that the ceiling and visibility were decreasing and asked, "Are you sure you want to give it a try?" According to the station manager, the captain replied that he would try the approach.

Several minutes later, the station manager went outside, saw that the weather was worsening, and estimated that the ceiling was about 200 feet, the visibility a half mile, and that fog was rolling over the top of the terminal building. He testified that this information wasn't passed on to the flight which was on its final approach and was thus in compliance with the regulations, which require that the weather be at minimums when an approach is initiated. A bit later he heard two small explosions and he was subsequently notified that the aircraft had crashed.

Surviving passengers reported that the flight was turbulent and that the crew had left the "seatbelts fastened" sign illuminated throughout the flight. One passenger noted that the aircraft's speed was reduced and that the aircraft "wobbled slightly" just before impact. He looked out the window and saw dense fog and then heard the sounds of impact with the trees.

The fog this evening was of the type where horizontal visibility fluctuates rapidly and the weather condition was such that turbulence was likely. There was also wind shear in the form of a diminishing headwind as the aircraft descended, which, when encountered, would tend to make the airplane want to nose down to maintain airspeed. Also, the aircraft was loaded forward of its approved forward center of gravity, or c.g., limit, which would require the pilot flying to use greater than normal force on the control column to counter any nose-down tendency in the wind shear encounter. The c.g. condition didn't affect the controllability of the airplane, just the force required to effect control.

In its report, the NTSB found that "the pilots probably saw the airport and the first officer began the circling maneuver for alignment with Runway 19 near the expected position—about 1.5 miles northeast of the threshold for Runway 23 (which had no lights) and at an altitude near MDA."

With the airport in sight, the descent below minimum descent altitude was begun and a turn started to align the airplane with Runway 19. And as the airplane entered the wind shear, the board found it likely that fog conditions were encountered, and that the pilots probably lost all visual reference that would have provided altitude information shortly before the aircraft struck the trees.

## Fog Aside

Fog aside, any pilot flying a night circling approach is subjected to some pretty strong visual illusions. When flying over a dark area and looking at lights in the distance, the illusion is one of flying at a higher altitude then is actually the case. You can often see pilots who haven't made a lot of night approaches affected by this. They tend to fixate on the lights in the distance, not look at the altimeter, and start a premature descent. One dark night

I was flying with a relatively experienced pilot, a flight instructor, who started descending out of 500 feet about a mile and a half from the airport on a circling approach. When I asked him what he was thinking about, he replied, "I don't know." I didn't know either and told him to restore power and level the airplane for a while.

## Fog Back

Fog is very insidious in what it does, too. Most pilots think in terms of being able to see progressively more as they descend, where with fog you often see progressively less. There are times when you can fly over an airport, look down, and see the runway or runways. However, the low-lying fog can preclude a successful landing because when the airplane is positioned on final approach you are looking through a lot more fog toward the runway than when flying directly above it.

## Blind

I was approaching the airport at Harlingen, Texas, early one morning. Another pilot was in the right seat of my airplane, and when we were five miles from the airport, we could see the runway and its lights. Continue. Descending on the glideslope, the runway was still in sight. Half a mile from the runway and it was still in sight. Then, as we passed over the end of the runway, starting the landing maneuver, the runway disappeared. I started to begin a missed approach, but before I could start, the runway came barely back in sight and I completed the landing that was about to complete itself anyway. After landing, we rolled into conditions with virtually unlimited visibility. Fog like that can be tricky and when it exists, a pilot has very little time in which to make decisions.

## Circling Figures

In operating turbine-powered airplanes (both prop and jet) for business transportation, the corporations of the U.S. have a good safety record, but the approach to an airport that doesn't have a precision approach system is a true chink in the armor. Seventy-five percent of the weather-related fatal accidents in a recent year occurred during such approaches, and a key item in corporate flying risk management is found in avoiding nonprecision approaches in minimum conditions, or visual approaches in marginal conditions—especially at night.

## Off the End

In larger jet airliners the more common result of problems with an approach is a trip off the other end of the runway. These usually are not major catastrophes, but they can illustrate the requirement for precision in all areas of flying. Often as not there is a contributing factor, such as a wet or an icy runway, and a wind shear effect on airspeed is common. A DC-8 that overran the runway on a breezy and rainy day illustrates how adverse factors can gang up on a flightcrew.

The aircraft was bound from Kennedy Airport in New York to Chambers Field, the Naval Air Station at Norfolk, where it was to pick up a load of military cargo and then fly to Iceland. In the package prepared for the crew by the dispatcher, the forecast for Norfolk called for a 500-foot ceiling and five miles' visibility in light rain and fog. The forecast wind was from 020 degrees at 10 knots. The dispatch package indicated that the preferred runway for landing would be 28, and that it was wet. According to the captain, there was nothing in the package that indicated poor braking conditions on the runway. He checked and noted that he could accept a tailwind component of four to six knots, based on the anticipated landing weight of the aircraft when it reached Norfolk.

As the aircraft approached the Norfolk area, the controller advised the crew that Chambers had scattered clouds at 200 feet, measured ceiling 500 overcast, visibility five miles in light rain-showers and fog. The wind was from 020 degrees at 13 knots, and while Chambers was not reporting any gusts, the controller advised the crew that at Norfolk International Airport, five miles away, gusts were being reported. No information was available to the crew about the braking conditions or the condition of the runway surface because no reports had been made to the controllers in the tower at Chambers Field.

The plan was to vector the flight for a precision approach radar (PAR) procedure to Runway 10. (Unlike the ILS system, which gives steering commands to the pilot through cockpit instrumentation, the PAR is operated from the ground, with a controller giving the pilot steering and rate-of-descent commands. This system is widely used in the military.)

As the aircraft was maneuvering for its approach, the controller told the crew to "expect a heavy wind shear two to one miles from touchdown, had a heavy 141 report it when coming in last." (A C-141 is a large four-engine military transport. The wind shear in this case would be related to a change in wind direction or velocity.)

The captain instructed the first officer to monitor the ground-speed-to-true airspeed differential, using the aircraft's inertial navigation system, in order to anticipate wind shear effects. He added 10 knots to his reference speed for the approach, to compensate for wind shear.

As the aircraft proceeded toward the runway, lined up with it, the first officer reported that the headwind was "30 knots on the nose." The surface wind at Chambers Field was given as 360 degrees at 20, or 100 degrees off the nose on the left side. This information indicated that the aircraft would experience a decreasing headwind of 30 knots as it descended through the varying wind—a situation that would make prudent the use of extra airspeed on final. When the aircraft was two miles from the runway the first officer reported that he had the runway in sight. At this time, the headwind component was 15 knots and the wind was shifting to the left of the nose of the aircraft. As the aircraft continued, the controller said, "Decision height, you are on glidepath, on course."

A crewmember recalled that the ground speed of the aircraft was 161 knots at the threshold of the runway and that there were a few knots of tailwind. The captain said that with a wind of 360 degrees at 20 knots, he expected a slight tailwind at touchdown.

The aircraft touched, and, according to the captain, skipped back into the air. The skip was "measured not in feet but in inches."

The captain stated that the airplane touched again within the first 3,000 feet of the 8,068-foot-long runway. After this touchdown the spoilers on the wing deployed, the captain selected reverse thrust and immediately applied the brakes. The captain stated that the brakes were ineffective; the first officer stated that he began to be concerned with the ability to stop the airplane when he saw the marker alongside the runway indicating that 4,000 feet remained.

Then the first officer reported, "You got 3,000 feet left." The captain recalled that he had full braking applied at this time but the airplane was not slowing down. The captain then told the first officer to get on the brakes with him and both crewmembers said that the brakes were fully applied but were ineffective, and that they never had control of the forward velocity of the airplane.

As the airplane passed the 2,000-foot marker on the runway, it drifted off the right side. It slowed slightly in the mud. However, it was headed toward a car stopped on a road at the end of the runway, so the captain steered it left, back toward the runway. The aircraft subsequently crossed the road and stopped.

Three firefighters said that they saw the airplane touch down abeam of the control tower, that they heard the reverse thrust, and saw spray from behind the airplane. The lieutenant in charge of the crash-fire-rescue truck ordered the vehicle to respond while the airplane was still moving down the runway because he believed the airplane would not be able to stop on the runway.

The motorist in the car near the end of the runway said that he first saw the airplane when it was 400 to 600 feet from the end of the runway. He said that he realized that it wouldn't be able to stop on the runway, that water "was flying up under the wings," and that the airplane appeared to be "unstable in that the wings were slowly dipping somewhat from side to side." He said that as the airplane moved closer to his car, "The wheels of the aircraft appeared to be locked and . . . the aircraft appeared to be sliding." The motorist said that he put his car in reverse to move away from the airplane and that it left the runway and crossed the road ahead of him with the right wing of the aircraft passing over the hood of his car.

A group of Navy personnel inspected the runway within 15 minutes and the consensus was that the runway was wet with numerous patches of standing water estimated to be ½ to ¾ of an inch deep. The Aviation Safety Officer said that "approximately ½-inch to ¾-inch of water was on the center and south side of Runway 10 in the last 3,000 feet. Water runoff to the north side was prevented by a strong wind even though the runway is graded. This situation is routine at Chambers Field and causes water to pool rather than drain even long after the rain has stopped." The driver of the fire truck stated that there was so much water on the runway that he was concerned about losing control of his truck as he drove down the runway. None of the air traffic controllers in the tower reported that they were aware of the tendency of the runway to flood under certain meteorological conditions.

**THE INVESTIGATION**

The National Transportation Safety Board's investigation of this accident showed that the airplane was in the correct position to complete the landing, but it also showed that the airplane's indicated airspeed was 10 knots above the reference speed at this point and that the airplane landed as much as 3,800 feet down the runway. The ineffectiveness of braking was attributed to the flooded runway and the manner in which the airplane was landed.

The NTSB found that the runway conditions at the airport had existed for at least 27 minutes before the accident but were not transmitted to the flightcrew or to the final approach controller.

A lot of factors were covered in the NTSB report on this accident, including an error in the dispatch package for the airplane that listed a higher than actually allowable maximum landing weight for the conditions that existed, the failure of the first officer to report the airspeed as 10 knots too high when the airplane was 200 to 300 feet high, the fact that reverse thrust was not used to the maximum degree possible, and hydroplaning (where water forms a film between the tire and the pavement, precluding any braking effectiveness). The probable cause was laid to the flightcrew, with the failure of airport management to disseminate hazardous runway warnings and the failure of the air traffic controllers to inform the crew that there was water on the runway. One board member suggested an alternative probable cause with first emphasis on a lack of communication between the tower and the crew on runway conditions, as well as the condition itself.

### DOMINOES

That accident shows how a crew that was basically doing a good job can get caught. They were told about wind shear, so they correctly monitored speeds and made allowances for the type of shear that appeared most likely. The aircraft was positioned properly for the landing, with the only factor at that point being a bit of extra airspeed. That was the extra that was being carried for the shear; when the effect of the shear failed to dissipate the extra airspeed, the crew should have done so. With a wet runway and a crosswind that was adding a little tailwind component to the proceedings, the runway length was marginal and a best effort would be required to stop.

---

## *There but for the Grace . . .*

Any pilot who has flown much has encountered a situation similar to the one faced by this DC-8 crew. I recall an approach to a small airport in Ohio when many of the same factors came into play.

The weather was bad, with minimum ceiling and visibility for the nonprecision approach to the airport. I would have to circle for a landing on the 3,500-foot-long runway. Using the air carrier safety requirement of having to be able to land and stop in 60 percent of the available runway, I was fat with pavement. I had the margin required by that rule plus an extra 500 feet.

The circle was kept close to the airport with the result that when I turned on final approach, the airplane was both higher than a normal approach slope, and the airspeed was a little faster. At an altitude of 500 feet there were many decisions to be made. How much would the extra speed erode the runway margins? Would I be able to get the airplane down by the time it reached the runway? Finally, because all this started with a runway that was substantially longer than required, I hadn't given much thought to whether the runway was wet or dry or whether or not there was a slope that would introduce a downhill component to the landing. The decision was to continue, and it worked okay but with heavy braking and use of most of a runway that had, a few moments before, appeared to be a piece of cake.

When runways are close to the minimum length, a lot of things have to be done precisely and decisions have to be made on a continuous basis as the airplane nears the runway. There's a discipline that must be enforced here, too. Some would read the account of the DC-8 and wonder why the captain did not elect to go around when it became obvious that the touchdown was well down the runway. That is really a split-second decision and one that can be fraught with peril. Usually, the consequences of sliding off the end of a runway are not fatal. The occupants are passengers in a machine that is decelerating and has the potential of coming to a gentle stop. Wait until too late, try a go-around, and if something is struck it will be in an airplane that is going much faster and that might be accelerating.

## Long Landing and Flukey Winds

Flukey, for lack of a better word, winds can affect a landing and these are often found when the wind is blowing over an uneven surface and toward a runway. At the huge air carrier airports, such as Dulles near Washington, this isn't a consideration, but at other airports, it can have a pronounced effect on aircraft operations. The example here happened in fine weather but on a runway that was minimum in length.

At the time, the runway at Harry S. Truman Airport in the Virgin Islands was 4,658 feet long with a 500-foot "overrun" that increased the effective landing length of the runway to 5,158 feet. That is shorter than most runways used by Boeing 727 aircraft; however, given the weight at which the aircraft were flown into and out of the field, the runway met all the requirements for air carrier operations. There were some restrictions placed on use of the airport by the airline's Boeing 727-100 aircraft, including a requirement to use Runway 9 only for landing (because of terrain), all landings to be made by the captain, and a maximum tailwind component of four knots on a wet runway and six knots on a dry runway.

The surface wind was from 120 degrees at 12 to 14 knots and the captain said that the approach to Runway 9 was started at reference speed plus 20 knots, and this was reduced to a speed 10 or 15 knots above reference speed for the approach. The captain said that the airplane was just a "shade below" the glide-slope in the area of the runway threshold and the lowest speed the captain could remember seeing was the reference plus 10 knots. The captain stated that he retarded the power gradually and then, when the landing was assured, brought the power back to idle. He said that the aircraft was aligned with the runway and he felt comfortable as he began the flare for landing.

### TURBULENCE

The captain said that he did not anticipate the turbulence that was encountered shortly after the landing maneuver started. This caused the right wing to drop and after the captain leveled the wings the first officer told him that the aircraft was high. The captain said that the turbulence seemed to buoy the airplane; however, after the first officer's call of being high, he "got it on the ground." The first officer estimated that the aircraft landed about 2,200 or 2,300 feet down the runway and that he wasn't worried about the length of the landing.

### GO AROUND

The captain said that he decided just before touchdown that the aircraft could not be stopped in the remaining runway and that, almost simultaneously with touchdown, he called for a go-around and moved the power levers forward. The cockpit voice recorder disclosed that when the captain called for the go-around he did not order a change of flaps setting. The first officer then asked the captain if he wanted the flaps set at 25 degrees: the captain responded "flaps fifteen," and the first officer later stated that 25 was the correct setting and rather than debate the point he just set the flaps there.

The captain said that after he advanced the power he couldn't see an indication on the engine instruments that the power was increasing to the prescribed value. He said there was no sensation of power being applied or of the aircraft accelerating so he pulled the power levers back and applied the wheel brakes. The aircraft continued across the 500-foot overrun and hit the ILS localizer antenna and a fence. The right wingtip then struck an embankment and the outboard portion of the wing was severed. The aircraft crossed a road, destroying several automobiles, and stopped in a gas station and against a rum warehouse. A fire broke out as the airplane slid to a stop; 35 passengers and two flight attendants died.

The NTSB determined "that the probable cause of this accident was the captain's actions and his judgment in initiating a go-around maneuver with insufficient runway remaining after a long touchdown. The long touchdown is attributed to a deviation from prescribed landing techniques and an encounter with an adverse wind condition, common at the airport.

"The nonavailability of information about the aircraft's go-around performance capabilities may have been a factor in the captain's abortive attempt to go around after a long landing."

In its findings, the safety board found that the captain did not follow company procedures calling for the use of a 40-degree flaps setting for landing (he used 30 degrees) which increased the landing roll, provided lower drag, lessened the decelerative capability of the aircraft, and made it more susceptible to the atmospheric or aerodynamic factors that would cause it to float down the runway. They also found that while the captain did realize that the remaining runway was critical with regard to stopping the aircraft, he did not know that the remaining runway was even more critical with regard to the execution of a go-around.

## TIME

From the time the aircraft reached a height of 10 feet above the runway until the time the captain said "let's go around," only 10.5 seconds elapsed. From that time until the sound of the impact took an additional 13 seconds. The board's findings suggested that a successful go-around was possible immediately at the onset of the float, after the wing dropped in the turbulence, and probably after the wings were leveled. This was a period of about four or five seconds or a little more. When the airplane touched, the board found that it could have been stopped within the confines of the runway but the go-around the captain elected to attempt was impossible.

This pilot was experienced in using this particular runway, so the following doesn't apply to this accident but it is a factor in some others. Not many pilots fly aircraft off runways which are minimum in length for the aircraft and conditions. Airports and airplanes are such that we usually have far more concrete than is required. Indeed, the rules require that an air carrier aircraft be able to land and stop within 60 percent of the runway available for the aircraft to be allowed to use that runway. That is a good margin. Yet often when riding on an airliner, the turnoff is at the end of the runway. At times that is because the crew doesn't make a maximum effort stop; at other times it might be because the speed was a little bit on the high side. But when faced with a runway that is just barely long enough for the operation (with the required safety margin) it becomes a precise business, as you can see from the DC-8 and 727 accidents outlined here. A changing or shifting wind can play a big part, and when variable winds start affecting the airplane before or during the landing maneuver, the pilot's decision-making process does not enjoy the luxury of a lot of time. A landing, especially in a shifting wind condition, is far from being as exact a maneuver as, for example, a takeoff.

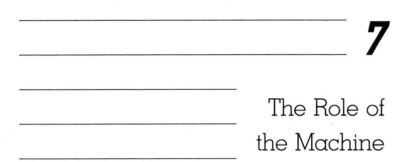

# 7

# The Role of
# the Machine

IF THERE IS an area where the forces of regulation have been effective in promoting aviation safety, it is in the standards by which airplanes are certified. It's not perfect, nothing is, but the system by which airplanes are designed and put into service is excellent. When the number of new types is considered, the complexity of modern aircraft is acknowledged, and the new areas that have been pioneered are added, the record of new aircraft has to be admired. This didn't come about without a little pain. Some tough lessons were learned after World War II, when a lot of new designs were put into service. The British Comet, the first jet airliner, was probably the most notable example of a problem encountered when moving to faster, higher flying airplanes. The aircraft had a skin cracking problem around the windows, which propagated when the airplane was pressurized and resulted in a catastrophic rupture of the fuselage. The Boeing 707, the first jetliner from a U.S. manufacturer, had a smoother entry into service, though a number of modifications were made to the 707 over the years—results of lessons learned in actual service.

Some of the newest airplanes have "glass cockpits," meaning that all those little dials and gauges are replaced, in whole or in part, with cathode ray tube displays. Every pilot has said, when first looking at a glass cockpit, "What happens if all the lights go out?" So far it hasn't happened, and the smoothness with which the new glass cockpits have been integrated into service is firm evidence of thoroughness in certifying new systems.

## All Different

But all airplanes are not alike, and as a result, the accident rate in them varies. This, though, has more to do with the relationship between the airplane and the pilot than with any design feature or characteristic of the airplane. This is least noticeable in airline aircraft with more than 30 seats. Here the standards of both certification, pilot training, and operation are greatly standardized, meaning that less variation is likely in the safety record of individual airplanes. In the history of large airplanes operated by major airlines in fleets of substantial size, there is little history of a specific type having major problems. Even the DC-10, which was grounded for a while after well-publicized problems, has, on balance, a good accident record.

## Chicago

The most famous DC-10 accident in the U.S. involved the airplane that literally lost an engine off its left wing on takeoff in Chicago. Following the accident and the inspection of other DC-10 aircraft, the FAA suspended the airworthiness certificate of the aircraft. The NTSB's probable cause of this accident is noteworthy because it covers a lot of territory:

"The National Transportation Safety Board determines that the probable cause of this accident was the asymmetrical stall and the ensuing roll of the aircraft because of the uncommanded retraction of the left wing outboard leading edge slats and the loss of stall warning and slat disagreement indication systems resulting from maintenance-induced damage leading to the separation of the No. 1 engine and pylon assembly at a critical point during takeoff. The separation resulted from damage by improper maintenance procedures which led to failure of the pylon structure.

"Contributing to the cause of the accident were the vulnerability of the design of the pylon attach points to maintenance damage; the vulnerability of the design of the leading edge slat system to the damage which produced asymmetry; deficiencies in the Federal Aviation Administration surveillance and reporting systems, which failed to detect and prevent the use of improper maintenance procedures; deficiencies in the practices and communications among the operators, the manufacturer, and the FAA, which failed to determine and disseminate the particulars regarding previous maintenance damage incidents; and the intolerance of prescribed operational procedures to this unique emergency."

On the fateful takeoff, the engine thrust was stabilized when the aircraft had reached an airspeed of 80 knots. The proper speed callouts were made at the time the aircraft passed through V1, the speed above which the takeoff would continue after an engine failure, and Vr, the speed at which the pilot would rotate the aircraft in the pitch axis to the proper attitude for initial climb. The last stable takeoff thrust rating on the Number 1 engine was recorded two seconds before lift-off. One second later the word "damn" was recorded on the cockpit voice recorder, which then ceased operating.

The aircraft became airborne about 6,000 feet from the point where the takeoff run started. The flight maintained a steady climb rate of 1,150 feet per minute on two engines. The engine that had been mounted on the left wing had separated from the aircraft at about the time of lift-off. During the initial portion of the climb, the aircraft was flying at an indicated airspeed of 172 knots. The nose-up pitch attitude of the aircraft was 14 degrees, which was what the flight director would have commanded for a two-engine climb. At this pitch attitude, the airspeed started decreasing and when it reached 159 knots the aircraft started rolling to the left despite control inputs to the contrary by the flightcrew. The DC-10 rolled past vertical, struck the ground, and disintegrated.

### COULD HAVE

When the engine tore loose from the airplane it damaged the devices on the leading edge of the left wing that create lift for takeoff and landing. When these devices retracted because of damage and a loss of hydraulic pressure, the stalling speed of the left wing increased. The high-lift devices were still extended on the right wing so the stalling speed of that wing remained lower. As the flightcrew flew the airplane at the prescribed pitch attitude the airspeed fell below the 159-knot stalling speed of the left wing and thus control was lost.

Thirteen pilots flew the flight profile of the accident in a simulator and in every case, control was lost when the pilots attempted to track the flight director's pitch command bars. In many cases, when the aircraft started to roll at the 14-degree nose-up attitude, the pilots lowered the nose, increased airspeed, and continued flight. But if they attempted to regain the 14-degree nose-up, the aircraft would start to roll again.

Based on the likely damage to the aircraft when the left engine tore away, pilots and test pilots who testified at the NTSB hearing believed that both the stall warning system on the aircraft and the slat disagreement warning light were inoperative. They stated that the flightcrew can't see the left engine and wing from the cockpit and therefore had no warning until the roll to the left commenced. Under these circumstances, none of those testifying felt it reasonable to expect the flightcrew to react in the same manner as those pilots who were able to recover in the simulator. The simulator pilots knew what was coming; the pilots flying the actual aircraft had no way of knowing the nature of the damage to their aircraft.

**MAINTENANCE**

In the course of investigating this accident, the NTSB found that the airline had developed a procedure, when removing an engine from the aircraft, of raising and lowering the engine and pylon (which is used to attach the engine to the wing) assembly as a single unit, using a forklift-type supporting device. The DC-10 field service representative was contacted by airline service personnel about this procedure and conveyed word on the procedure to his superiors. According to him, the manufacturer would not encourage this procedure due to the element of risk involved in the remating of the combined engine and pylon assembly to the wing attach points. The airline was so advised.

Because of a modification, there had been substantial activity in removing and replacing the wing-mounted engines of DC-10s in the fleet. Some airlines were using forklifts to remove engines and pylon as one unit; some were removing the items separately using an overhead crane. A total of 175 procedures had been accomplished at the time of the NTSB investigation, and in a required inspection of the DC-10 fleet, nine cases of damage and cracking were found. In all nine cases the engines and pylons had been changed with a forklift.

If the cracks were caused by a maintenance procedure, why was the DC-10 grounded after the accident? One of the rules under which the airplane was certified states: "The motion of the flaps on opposite sides of the plane of symmetry must be synchronized unless the aircraft has safe characteristics with the flaps retracted on one side and extended on the other." Further, a special condition requires that the aircraft be shown to be capable of continued flight and landing after "any combination of failures not shown to be extremely improbable." According to FAA witnesses at the DC-10 accident hearing, "extremely improbable" is considered to be one chance in a billion.

When the engine separated from the aircraft, the accessories driven by the engine were apparently also lost and the hydraulic pressure and supply lines connecting the hydraulic pumps with the system were severed. The hydraulic system would then lose all its fluid and hydraulic pressure would not be recoverable using other systems. The separation of the engine also severed the electrical wire bundles inside the pylon. This would cause the loss of power on one of the electrical buses, in this case the one that operated a number of instruments, the cockpit voice recorder, and, most importantly, the stall warning and slat disagreement warning light systems. The flightcrew might have been able to restore power to these items but the NTSB believed that they did not try, either because of the nature of the overall emergency involving other systems, which they probably perceived to be more critical than the electrical problems, or because the time interval did not permit them to evaluate and respond to the indicated electrical emergency.

## REDUNDANCY

Because of required redundancy in electrical and hydraulic systems, the loss of systems powered by the left engine shouldn't have affected the crew's ability to control the aircraft. However, when the engine and pylon separated, four other hydraulic lines and two cables were severed. This caused the loss of use of spoiler panels on both wings, used to control the aircraft in the roll axis, but the primary problem was in removing hydraulic pressure from the leading edge slat system, allowing air loads to retract the left outboard slats.

The NTSB found the evidence conclusive that the aircraft was being flown in accordance with prescribed engine failure procedures. The consistent 14-degree pitch attitude of the aircraft indicated that the commands of the flight director were being followed, and, since the captain's flight director was inoperative due to the loss of the one electrical bus, this confirmed that the first officer was flying the aircraft. The configuration of the aircraft was such that there was little or no warning as the outboard section of the left wing reached its stalling speed. Since the roll to the left started when the aircraft was flying at a speed six knots above that which was correct for an engine out climb (with all the slats extended) the crew probably did not suspect that the roll to the left was associated with a stall. Also, the stickshaker, which warns of a stall, did not activate.

In its findings, the NTSB included the fact that the design of the stall warning system lacked redundancy and crossover information to the left and right stall warning computers; that the design of the leading edge slat system did not include positive mechanical locking devices to prevent movement of the slats by external loads; and that at the time the airplane was certified, the structural separation of an engine was not considered and multiple failures of other systems resulting from this single event were not considered. The FAA subsequently grounded the DC-10 for over a month, and when the aircraft was returned to service modifications and inspections were required and the procedure of removing the engine and pylon as one assembly was specifically prohibited.

## Monitoring

The FAA, the service technicians, aircraft manufacturers, and pilots have a good system of communicating with one another about problems with airplanes. If something is breaking, or if a significant maintenance history is developing on an airplane, word travels fast. Then the FAA and the manufacturer have to decide whether the problem is something that must be addressed on all aircraft, engines, or accessories of the same type. If this is decided, then basically they have decided that the airplane, as built and being flown, no longer meets the required standards and must be modified. An airworthiness directive is issued. At times, the requirement is to fix or change something immediately, grounding the airplane until the change is made. In less severe cases, the requirement is to fix or change something within a specified period of time. In the case of the DC-10, the system suffered a lapse. But it is still a good system.

Occasionally, an airworthiness directive is issued that requires a change in the way an airplane is operated, based on accident history. The P210, a piston-powered single built by Cessna, had in its early years an almost continuous string of airworthiness directives on the engine. It was a temperamental powerplant that actually had little wrong with it that couldn't be handled from the pilot's seat. But because of often inaccurate engine instrumentation, there were numerous cases where an engine failure resulted from the manner in which the pilot operated the engine. Some new aircraft hardly got out of sight of the factory before the engine failed. The result was a number of modifications that made the engine less sensitive to damage from being run with the fuel/air mixture too lean. This cost performance in the airplane, which seemed unfair to those who understood it and had had no problems, but that's the price of having to make airplanes as safe as possible for all who operate them.

## Instruments

Another significant light airplane change that came about because of a history of accidents related to the power source for flight instruments. Most are driven by either a vacuum or pressure air pump, and for years single-engine airplanes flew with one pump and there was no established history of trouble with pumps. Then, as the pumps started being used for other duties, such as operating deice boots on the wings and tail, and airplanes started operating at higher altitudes where pumps had to work harder, a pattern of pump failures developed. And even though aircraft have alternate, electrically driven instruments to use, flying with these is both difficult and something that pilots didn't practice a lot. As a result, control of the aircraft was lost and a crash followed. So the FAA instituted a requirement for either a backup power source for instruments or redundant electric and vacuum instrumentation on some airplanes. Aircraft manufacturers started incorporating dual or backup air systems in all aircraft built and approved for instrument flight rules flight. The FAA also put emphasis on the testing of pilots on the art of flying with less than a full panel of instruments.

## Instrument Loss

When the instruments in an airplane go awry, even a professional flightcrew can have real problems. The cockpit conversation on the flight deck of a cargo flight illustrates this.

CAPTAIN: Gyro's, ah, screwed up.

FIRST OFFICER: Yeah, I keep turnin' it was the other way.

CAPTAIN: Yeah.

FIRST OFFICER: Wings level now. . . . Could you switch it over to number one?

FLIGHT ENGINEER: Yeah.

FIRST OFFICER: Yeah, it's level now.

FLIGHT ENGINEER: Better?

FIRST OFFICER: Yeah.

FLIGHT ENGINEER: Okay.

FIRST OFFICER: Cruise . . . might be just, ah . . . horsepower reads low on number two . . . horsepower readin' low.

CAPTAIN: Yeah.

FIRST OFFICER: Temperature's readin' high and the fuel flow is normal.

FIRST OFFICER: Are you set . . .

CAPTAIN: They already changed the, ah, anti-ice valve.

FLIGHT ENGINEER: Ah, wait a minute . . . reset the fuel flow to around eighty . . . and you get a hot temp . . . might just be the horsepower calibration.

FIRST OFFICER: Did he say tops is this area, is that what he said?

CAPTAIN: It's clear above and below.

FIRST OFFICER: What do you figure?

CAPTAIN: I don't know.

FIRST OFFICER: Ah about seventeen fifteen.
CAPTAIN: Yeah.
FLIGHT ENGINEER: Times and temps.
[*Sound similar to paper pages being turned.*]
FIRST OFFICER: What's happening here?
[*Sound of increasing wind noise.*]
FIRST OFFICER: You got it?
CAPTAIN: No.
[*Sound of landing gear warning horn starts, followed by sound of over-
    speed warning and then by a sound similar to structural failure
    during an inflight breakup.*]
UNIDENTIFIED VOICE: We're dead.

That was the last of the recording. The aircraft broke up in a right descending spiral and the wreckage was scattered over an area two miles long by one mile wide.

The NTSB found from the cockpit voice recorder that the crew experienced a problem with the Number 2 vertical gyro system (that drives the flight instruments) and selected the Number 1 system. Then, from the report: "Although the precise reason for the loss of control was not identified, an undetermined failure of a component in the vertical gyro system, perhaps involving the amplifier and associated circuitry, probably contributed to the cause of the accident by incorrectly processing data to the copilot's approach horizon."

This airplane was a four-engine turboprop and both the captain and the first officer had a full set of flight instruments. However, consider that if you have two instruments and one is giving correct information and one is giving incorrect information, the job of deciding which is right and which is wrong is not a simple one. In jet airliners, there is a third attitude indicator, so there you can take the best two out of three. But this third instrument is not required in turboprop airplanes and the FAA had rejected NTSB recommendations that the third instrument be required in turboprops, "due to the lack of flight control or electrical problems associated with this type of aircraft." There have, however, been two catastrophic accidents that have been related to instrument problems so it would appear that the NTSB has been correct in its recommendation.

## Engines

Often, even usually, when a single-engine airplane crashes the press reports will mention that the engine spluttered, or raced, or did something strange before the accident. The result of this is a public perception that there is something inherently hazardous about having but one engine on an airplane. The FAA feeds this with restrictions on the use of single-engine airplanes for passenger carrying in air-taxi flying.

This is unfortunate, for several reasons. One is quite obvious. Anybody can look at an airplane with one engine and tell you what will happen if that engine stops running. The airplane will come down. That there is a risk of this is patently obvious. So might it not be reasonable to suggest that anyone voluntarily riding in an airplane with one engine has to, by the simple appearance of the machine, be aware of any risk?

This becomes doubly important when viewing an airplane with more than one engine. The suggestion to an observer becomes one of redundant power. If it has two engines and but one of them fails to operate, couldn't the flight continue normally?

There are a lot of years of history to study on this, and the simple fact has always been and remains, if you are in a twin-engine piston airplane your chances of being killed after one engine fails are far greater than are your chances of being killed in a single-engine airplane after *the* engine fails.

Looked at another way, the total fatal accident rate per 100,000 flying hours is roughly equal on single- and twin-engine piston-powered aircraft that weigh less than 12,500 pounds. And where nine percent of the fatal wrecks in single-engine airplanes are related to an engine failure, 24 percent of the fatal accidents in twin-engine airplanes are related to engine failures. And where virtually all twin-engine airplanes are factory-built and certified airplanes, the singles include homebuilts that may or may not use a certified engine.

Shining light on this from a different direction, about six percent of the accidents resulting from an engine failure in a single-engine airplane result in fatalities, where 23 percent of the accidents following engine failure in a twin-engine airplane are fatal. Further, if you are looking at all accidents, as opposed to just fatal accidents, it is true that "engine failure or malfunction" is the most prevalent type of accident in single-engine airplanes, accounting for 25.5 percent of the total, but engine failure or malfunction can also be the most prevalent type accident in small twin-engine airplanes. There, in a recent year, this accounted for a whopping 33.9 percent of the accidents. No matter how it's sliced, the small twin fares much worse after the failure of one engine than does the single after a total power loss.

Looking further, in an average year there might be about 70 fatal accidents in singles and twins with piston engines that are related to engine failure. Of these, less than 20 involve some established mechanical cause for the failure, such as a valve or cylinder or crankshaft breaking. Another 20 will have failed for "undetermined" reasons, which means that nothing could be found to indicate why the engine failed to run. The rest involve a clear case of the pilot running out of fuel, or mismanaging the fuel supply—trying to fly with the fuel tank selector positioned for a tank with no fuel when there was fuel in the other tanks, for example.

## Why?

Why is it that something that seems obvious—that the single-engine airplane would have more serious accidents than the twin following an engine failure—is actually the reverse of the truth? The answers start appearing when you examine the FAA requirements for certifying single-engine airplanes and small twins.

To begin with, a single must have a stalling speed of 61 knots or less. That lowest speed at which an airplane can be kept under control is important because it directly relates to the forces that will be encountered in an accident. If, for example, an airplane flies into trees at 60 knots, the accident is far more survivable than if it does the same thing at 90 knots.

A twin must have a stalling speed of 61 knots or less, except this is waived if the twin can meet an engine-out climb requirement at 5,000 feet. This requirement is minimal—if the twin stalls at 80 knots the required climb is but about 172 feet per minute, meaning that for every minute of flight the airplane ascends 172 feet. This is far shallower than even a normal approach angle, for example, and can be reduced by higher than standard temperatures or turbulence. But for competitive reasons, virtually all manufacturers of small twins design their airplanes so that they just meet the minimum requirements. That way, payload can be maximized, and if the FAA says this minimum is enough, then it is enough. Right? Wrong, if the lessons that have been learned from years of accidents in these airplanes are heeded.

When the power being put out by an airplane is asymmetrical, as it is on a twin flying on one engine (unless the engines are arranged fore and aft), the airplane is difficult to control, and if the speed drops below a certain value, known as the minimum control speed, then the aircraft becomes uncontrollable with one engine operating and the other not operating. A substantial number of twin accidents become unsurvivably serious because the pilot loses control of the aircraft through a combination of low airspeed and asymmetric thrust.

The pilot flying a twin has a very demanding task after an engine failure. There are decisions to be made and flawless flying technique to deliver. A pilot flying a single has fewer decisions to make and the flying technique required to land a single in a random off-airport location in a manner that is survivable is probably less demanding than handling a twin at a critical time with one engine out.

Then there is the matter of stalling speed, and the way the aircraft are built. Even if it is in control, a twin is going to hit harder in an off-airport landing than a single. The accident is less likely to be survivable because of speed. Also, in most singles, the heaviest and most dense object, the engine, is leading the way through the trees or whatever. In the twin, the engines are on the wing and they would tend to drag the lighter fuselage through whatever might present itself in an off-airport landing. In twin-engine airplanes, 82 percent of the controlled collisions with the ground result in fatal injuries. In singles, less than 60 percent of the controlled collisions with the ground result in fatal injuries. That hardly recognizes colliding with the ground as an exciting new outdoor sport in either type airplane, but it does illustrate how the requirement for a lower stalling speed in singles has done more to improve the safety record there than has the minimal single-engine climb requirement for twins.


## Dark Night, Engine Out

A twin-engine Beechcraft Baron 58 was to be operated on a night flight by an experienced pilot who had over 1,000 flying hours in this type aircraft. According to the NTSB report, the aircraft was estimated to be just under 100 pounds over the maximum allowable weight. It was a warm night with light winds.

Witnesses saw the aircraft roll down the runway and take off with no apparent problems. One witness said he last saw the aircraft at an altitude of 100 to 150 feet. He was momentarily diverted by another witness, but after hearing a popping sound he turned to look for the aircraft. Instead, he saw a ball of fire east of the airport.

The NTSB report said: "The evidence indicates that the left engine failed shortly after the aircraft took off because the crankshaft broke at the Number 7 crankcheek."

The right engine apparently continued to function normally. The landing gear of the aircraft remained extended until the aircraft struck the ground. The NTSB report stated that the pilot had little time or altitude in which to recognize the problem and respond with the appropriate corrective action.

In its section on aircraft performance, the board called attention to the fact that aircraft in this category are not required to be capable of climbing in the takeoff configuration with a critical engine inoperative and its propeller windmilling. The board compared this to transport category airplanes (basically those with a maximum takeoff weight in excess of 12,500 pounds) in which the takeoff weight must be limited to that at which the airplane does have climb capability with one engine out, its prop windmilling, and the landing gear extended. From here, a pilot would have the capability of improving things by feathering (aligning with the wind) the propeller blades on the failed engine and retracting the landing gear to reduce drag.

The NTSB's probable cause finding: "The National Transportation Safety Board determined that the probable cause of this accident was the sudden failure of the airplane's left engine at a point on the takeoff flightpath where the airplane's single-engine performance in the takeoff configuration and its height above the ground combined to make the pilot's ability to sustain flight marginal. The pilot's failure to retract the landing gear and control the airplane to maintain a safe airspeed contributed to the accident and were factors in causing the high acceleration loads when the airplane struck the ground."

## Hopeless?

Often when writing things like this, I have examined the motive for doing so. Someone once suggested that this sort of information could "kill" the light twin. That is certainly not the intention. The intent is to show what the actual risks have proven to be, based on the accident history. If a pilot avails himself of the best in initial and recurrent training, and if the airplane is flown off large airports at reduced weights, chances are the engine failure risk to life might be reduced to a level as low as it is in a single. It might even be argued that the risk in the twin could be made lower than in the single by following transport category requirements and standards. But on the other hand, the pilot of a single can manage risks, too. By buying the best maintenance, by making sure there is plenty of uncontaminated fuel on board, and by blowing the whistle at the first sign of abnormal engine operation, the pilot of a single can reduce risk. Neither pilot can completely eliminate the risk of having an accident because of an engine failure, but it can be kept much lower than average with a little effort.

## Other Failures

In light airplanes the failure of other systems has and does result in fatal accidents. The systems that power the instruments were mentioned. If an airplane loses electrical power, this can result in the loss of key items. Managing these risks is up to the individual, and besides knowing how to fly the airplane with less than full equipment, a careful pilot who oversees the maintenance of the airplane can avoid a lot of problems. The air pumps that are used, the alternator systems, the batteries, and the instruments all have finite lives. And where enhancing the safety of flight depends on a single system, it stands to reason that the key components of these systems should be replaced at specified intervals rather than wait for them to wear out and fail. I used to average at least a couple of systems failures a year on my airplane; a few years ago I started replacing the alternator and air pump once a year (about 500 hours of flying for me) and haven't had a failure since. That's not to say I have totally eliminated the risk from the failure of a system, but it does indicate that this is an effective way of reducing the risk.

## The Airframe

Occasionally a large airline aircraft suffers an in-flight failure of structure that leads to an accident. In Japan, the 747 that crashed in 1985 apparently lost a significant portion of its vertical tail as well as suffering damage to the control system of the aircraft. The DC-10 we've already examined. When these events occur, it is more likely related to something having been damaged, or something failing because of metal fatigue, than it is caused by factors such as turbulence or pilot control inputs putting too much load on the structure. There are a few cases of the latter happening—the cargo airplane that broke up in flight after an apparent instrument problem is one example, but airline airplanes are quite strong. I can recall only one airline aircraft that suffered a structural failure around a thunderstorm in recent times, and in that case the storm was extremely severe and the aircraft apparently got to the worst place at the worst time. There have been cases of crews losing control of airliners and the aircraft making a high-speed plunge with many of the design parameters exceeded—but the airplane held together despite the stresses.

## Small Aircraft

The same thing is not true of light airplanes. There are a number of airframe failures—probably about 50 in an average year—which makes this a substantial portion of the fatal accident picture.

Weather is a factor in the great majority of airframe failures. Either a pilot who is not trained in instrument flying inadvertently flies into cloud and loses control of the aircraft, or a pilot who is instrument rated flies into turbulent clouds that exceed either the pilot's ability to control the airplane or the strength of the basic structure of the airplane. Most of the airframe failures involve an airplane with a retractable landing gear because these airplanes are aerodynamically cleaner and build speed to a value in excess of the design speed of the aircraft more quickly than airplanes with fixed landing gear. In virtually every case of airframe failure, the NTSB finds that the design limits of the aircraft were exceeded.

It would appear that this is an area where aircraft design could help, but some factors make it questionable that anything other than better pilot training would make the record markedly better. In most cases, the airplane is relatively low—below 10,000 feet above the ground—when control is lost. In the spiral dive that ensues, a very high rate of descent develops. The airplane then quickly reaches a point where, regardless of pilot action, it is going to hit the ground. Indeed, a number of the airframe failures occur after the pilot has recovered from the upset, and are a result of control pressures applied when the airplane is flying at an airspeed higher than that for which it is designed. In other words, once the airplane was clear of clouds, the pilot saw the ground rushing toward him and made a maximum effort to pull the airplane out—and there simply wasn't enough room to perform the maneuver without overstressing the airframe. Every airplane has a structure that is stronger than the requirement, and in these the part that just meets the requirements is usually the first to fail. Thus a pattern develops, and on some aircraft, such as the V-tail Beech Bonanza, the pattern has been cited as reflecting a fatal flaw in the airplane. But that airplane's record is no worse than similar airplanes such as the Cessna 210 and the Piper Comanche and Lance, all of which have tail sections sporting the usual three pieces.

## Interface

When looking at any pattern that develops in airframe or engine failures, the interface between the airplane and the pilot becomes important. As time has passed and the cost of accident litigation has increased, the insurance companies have become far more involved in evaluating this interface. For example, because of the airframe failure history of retractables, some insurance companies simply won't insure those airplanes unless the pilot meets a rather high minimum flying time requirement and has an instrument rating. On other types, insurance companies put a high premium on the airplane simply because it has a worse than average accident record.

It's usually relatively easy, based on the hindsight that comes from studying accident reports, to put together the reasons for any pattern in accidents. The surprising thing is that the FAA does so little about this. Despite two accidents involving large turboprops with instrument problems, the FAA declined to require auxiliary instrumentation. Despite the overwhelming evidence that the accident record after engine failure is twins is far worse than it should be, the FAA has no specialized recurrent training requirement in these airplanes. Nor has it changed the requirements for certifying these airplanes. This applies to airframe failures as well. A pilot can, for example, train in and take the required biennial flight review in a simple two-place airplane, yet that same pilot is required only to have a VFR check-out in a high-performance retractable-gear airplane before flying away IFR in that airplane.


## Ground Folks

The interface that causes a problem can be on the ground as well as the flight deck, as the people on a trijet airliner learned on a day that turned out to be very lucky for all concerned.

The airplane was descending to land in the Bahamas when the low oil pressure light on the Number 2 (center) engine illuminated. The engine was shut down because of this warning, and the captain of the flight elected to return to Miami. While en route, the low oil pressure lights for the other two engines illuminated. At this time the oil quantity gauges for all three engines read zero and a crewmember informed the air traffic controller of that fact and added: "We believe it to be faulty indications since the chance of all three engines having zero oil pressure and zero quantity is almost nil." The flight engineer then called the maintenance department on the radio and inquired about a common electrical source that might affect the instrumentation to all three engines.

Then, when the airplane was 80 miles from Miami, the Number 3 engine failed. The flightcrew said that at this point they realized that the indications of zero oil pressure and quantity were correct and not the result of a gauge problem. With but one engine running, the aircraft entered a gradual descent and the flight engineer called the senior flight attendant to the flight deck and instructed her to prepare the cabin for ditching.

The flightcrew was attempting to start the Number 2 engine that had been shut down earlier but had not been successful when the Number 1 engine failed. At this time the airplane was about 12,000 feet above the ocean and the rate of descent increased from the 600 feet per minute it had been with one engine, to 1,600 feet per minute with no engines. In other words, they had seven and a half minutes to go. In that time, the airplane would have been able to glide about 30 or 35 miles but it was 55 miles from Miami. An auxiliary power unit was providing electrical power so the controls, instruments, and radios were all working.

The flight engineer announced, "Ditching is imminent," while the aircraft was still above 10,000 feet and the senior flight attendant assumed it was about to ditch and instructed the passengers to assume the brace position. A virtual fleet of airplanes, helicopters, and ships were alerted and started in the direction of the powerless jet, which had 172 people on board.

The captain again tried to start the Number 2 engine but wasn't successful. Then he tried 1 and 3 with no better luck. The ditching checklist was being run and the captain tried again on Number 2, this time with success. The aircraft was, at this time, 4,000 feet above the ocean. The descent was arrested at 3,000 feet and the aircraft climbed gradually to 3,900 feet on one engine as it flew the remaining 22 miles to Miami where it landed successfully and, according to the NTSB report, the passengers deplaned "normally."

There were some interesting moments in the cabin during this emergency. The flight attendants said that before departure many passengers did not watch the life vest donning demonstration, which is usual. When they were told to prepare for ditching they didn't know how much time they had to prepare so they did it as quickly as they could. When the "imminent" announcement was made, all assumed the brace position and remained there until shortly before the landing at Miami. Able-bodied passengers were stationed next to emergency exits to help with egress. According to the flight attendants, the passengers were generally close to panic with some unable to function and others screaming throughout the emergency. Many had problems with life vests.

The airline's procedures required the selection and briefing of able-bodied passengers to assist with crowd control, doors, and slides. It was reported that many male passengers turned down requests to help; others were not selected or refused because they had consumed too much alcohol. But volunteers were found for all doors except one, in a section where few people were seated. Some passengers cited inadequate responses to questions about postevacuation procedures and the expectations of whether the airplane would float or sink immediately. There was no way to know because an aircraft of this type had never ditched and one's imagination would have been as accurate as any information that might have been available.

### WHY THE FAILURE?

What caused the engines of this big airplane to fail? It was quite simple. The NTSB determined that the probable cause was "the omission of all O-ring seals on the master chip detector assemblies leading to the loss of lubrication and damage to the airplane's three engines as a result of the failure of mechanics to follow the established and proper procedures for the installation of master chip detectors in the engine lubrication system, the repeated failure of supervisory personnel to require mechanics to comply strictly with the prescribed installation procedures, and the failure of . . . management to assess adequately the significance of similar previous occurrences and to act effectively to institute corrective action." A simple omission of three O-rings almost caused a major accident.

## Twins Across the Water

An engine-related item that is becoming important to those who fly across the Atlantic is the use of twinjet airliners on some of these routes.

Where a person riding in a single-engine airplane can clearly see what the deal is before flight, a lot of airline travelers are not aware of what type airplane they are riding on, and virtually everyone has come to expect aircraft used on long overwater flights to be three- or four-engine models. But now there are twins on these routes. Suffer the failure of one engine and you have just joined the ranks of those who fly oceans in single-engine airplanes.

The decision to do this was based on the aircraft having redundant systems, extra equipment, and on the proven or expected reliability of the powerplants. All this is well and good, but in the first six months of operation over the Atlantic, one operator experienced four engine shutdowns and one significant loss of thrust, according to the president of the Air Line Pilots Association. An article in *Air Line Pilot* outlined the incidents:

"An engine overheated during a step climb. The flightcrew shut down the overheated engine and diverted to Bangor, Maine, to avoid marginal weather conditions at Gander, Newfoundland, the closest airport. The diversion required flying on one engine for approximately two hours.

"The same engine . . . overheated during another step climb. The engine was throttled back until it appeared to be operating normally. The incident occurred after a significant portion of the cruise phase of flight had elapsed, so the flight continued to its original destination.

"Loss of engine oil resulted in an inflight shutdown and single-engine diversion of approximately 60 minutes to Keflavik, Iceland. Problems with the ground ILS equipment forced the pilots to fly a single-engine nonprecision approach in marginal weather.

"Loss of engine oil forced a Paris-to-St. Louis flight to divert to Goose Bay, Labrador.

"Faulty instrument indications of low engine-oil quantity and high engine-oil temperature caused a Munich-to-New York flight to divert to Prestwick, Scotland."

The president of ALPA charged that "the inflight engine shutdown rate and subsequent diversions do not appear, in our view, to meet the safety objectives of FAA Advisory Circular 120-42, Extended Range Operation with Two-Engine Airplanes. The greater the inflight engine shutdown rate, the greater the potential for total thrust loss on both engines."

Certainly unless the shutdown rate becomes much lower than it was in the first six months, the likelihood of having a twin ditch in the Atlantic after the failure of both engines is more than a one-in-a-billion remote possibility.

### OTHER MECHANICAL PROBLEMS

The other incidents are what you would expect. Things on airplanes do break, and things do fall off. But the way aircraft are designed and built, and the method in which problems are used to develop improvements for all aircraft in the fleet, leads to a high degree of mechanical reliability. The most technically complex airplane in the airline fleet, the Concorde, is delayed by technical problems on but five percent of its flights and, complex as the airplane is, the landing gear is the most frequent culprit. (No problem getting it down; the retraction is the most complex part of the cycle.) And there is, in the record, plenty of evidence that training is excellent and that airline crews can handle emergencies.

### DOUBLE FAILURE

A widebody twinjet was descending toward Denver when the left engine surged and exceeded a temperature limitation. About 18 seconds later, the right engine surged and exceeded the same temperature limitation. The crew shut down both engines, declared that an emergency existed, and then went about the business of restarting the engines. They were successful in getting a restart about 15,000 feet, after the aircraft had lost more than 10,000 feet.

### GEAR PROBLEM

The passengers and crew aboard a Boeing 727 were enjoying a normal climb out up to the point where the aircraft reached an altitude of approximately 10,900 feet. There a loud bang was heard, followed by the illumination of the red landing gear warning lights tagged "doors" and "right gear." Following procedures to handle this condition, the first officer moved the landing gear from the "off" to the "up" position. After this, the flight engineer reported loss of fluid and pressure in both hydraulic systems. The primary flight controls reverted to manual operation and the climb was terminated.

The air traffic controllers were notified of the problem and the crew elected to stay at 11,000 feet while troubleshooting the system. The captain advised that they would be dumping more than 19,000 pounds (almost 3,000 gallons) of jet fuel, to lighten the ship for landing.

The crew couldn't determine the position of the right main landing gear through a visual inspection hole provided for this purpose. So the aircraft made a low-level fly-by of the airport where personnel could see that the nose and left main landing gear were retracted and their gear doors were closed, with the right main gear retracted but with its gear door open.

The crew reviewed all the checklists covering this type event as well as other limitations. After the review, the captain elected to lower all landing gear by using the emergency manual extension procedure. The flight engineer read the pertinent instructions located near each landing gear manual crank socket and, starting with the left main landing gear, commenced to put the wheels down. The left main gear apparently operated per the plan and the green light indicating that it was down illuminated. Moving to the right main landing gear, the crank rotated normally but when the procedure was completed, the red "unsafe" warning light for this gear remained illuminated. The procedure was repeated for the nose landing gear, which came down and its green light illuminated.

After trying other things in the cockpit that would verify an unsafe landing gear condition, the crew did another fly-by and was informed that the left and nose landing gear were down but that the right landing gear remained up.

Faced with landing the aircraft with two down and one up, the crew once again reviewed procedures, apparently decided that they had done all they could do, and prepared for an approach with the right main landing gear retracted and without the availability of the hydraulic boost system for the flight controls.

The crew requested a landing on Runway 9 right in order to

have the grass area on the south of the runway on the right side, the direction in which the aircraft would veer with no right main landing gear. The aircraft landed on the runway; as it slowed, the right wing dropped and contacted the ground, the aircraft turned about 45 degrees right, the two extended landing gears collapsed, and the aircraft skidded to a stop about 100 feet south of the runway.

The emergency equipment was at hand and quickly sprayed foam on the aircraft and started to assist with the evacuation which was begun within 10 seconds after the airplane stopped. The cockpit crew went out through a cockpit window and to the right forward emergency slide to assist the passengers. All 152 passengers aboard the aircraft were evacuated through four doors, using four slides; the evacuation of a full airplane was well co-ordinated and carried out expeditiously.

As for the loud bang that led to the landing gear problem, it was the explosive blowout of one tire on the right main landing gear while it was retracted in the wheel well. The explosion caused damage that resulted in the loss of both hydraulic systems and precluded the emergency manual extension of the right gear.

The crew's actions were correct and in accordance with the aircraft's operating manual. However, the NTSB recommended that the FAA require a training program addressed specifically to the recognition, assessment, options, and procedures to be followed after the explosion of a tire in a wheel well. There were other recommendations, including one to determine the feasibility of protecting hydraulic system lines in the wheel well during a tire explosion.

In perusing the incidents that are reported, it is obvious that almost everything that can happen does happen. That's why, in ground school training programs and simulator flying, learning how the systems on an aircraft work and practicing emergencies and emergency procedures are a big part of the action. Pilots flying general aviation airplanes with a maximum takeoff weight of less than 12,500 pounds aren't required to have this training but many avail themselves of it because the worst, and most potentially lethal, thing a pilot can face in the cockpit is a surprise or series of surprises. It's far better to have knowledge, and to have considered in advance how to deal with anything that might happen.

# 8

## Collisions

OF ALL THE TYPES of aircraft accidents, collisions probably strike fear more deeply into pilots and passengers than any other. I've heard grizzled veterans say that they have everything figured out and covered by training or regulation but the collision—to most that remains an enigma. In chapter 2 we explored the relationship between regulation and collisions and a bit about the air traffic control system; here we'll dig more deeply.

To begin, the chances of being involved in a midair collision are small. That's scant comfort to someone involved in one but it is true. There are not a lot of them and they occur on a random basis. Some pilots fly all the way through a career without ever having another airplane come uncomfortably close. (But in one case years ago, a pilot had two midair collisions in the same general area—the second one was his last.) Most pilots do their absolute best to make everything work for them in avoiding collisions and certainly none underestimates the undesirability of being involved in an aerial encounter.

## Where?

It stands to reason that the higher the density of air traffic, the greater the chance of collision. This has to be at least partly true, but it is also true that where there is the highest density traffic, there are more things to use as aids in avoiding collisions. There's better air traffic control radar service and there are procedures that a pilot can use to avoid areas of heaviest concentration, such as the final approach courses to busy airports.

To provide a framework, a little review on the way the air traffic control system works is needed. If an airplane is flying VFR, visual flight rules, the pilot is responsible for separating the aircraft from others. By calling an air traffic control facility, the pilot can often get advisories on other traffic in the area if the controller has time to give this service. Being in contact in no way relieves the pilot of the responsibility to see and avoid other traffic. VFR flying must be done clear of clouds, in relatively good weather. It is prohibited at and above 18,000 feet. For a VFR aircraft to enter an airport traffic area (within five miles of an airport with a control tower at an altitude below 3,000 feet above surface) the pilot must be in contact with that control tower. However, the control tower does not provide separation from other VFR traffic. A tower's primary job is to control the runway at the airport, keeping traffic there and on the ground moving in a safe and orderly manner. To fly into a terminal control area VFR, an airplane must have a clearance from air traffic control. Terminal Control Areas are established at the busiest airline airports and traffic is separated. To fly into an Airport Radar Service Area, a de facto TCA, a pilot must contact a radar controller before entering.

When flying at an airport without a control tower, there's a radio frequency for pilots to use in announcing position and for talking with other pilots in sort of a "self-operated" air traffic control scheme.

## Instrument Flight Rules

When an airplane is flying on instrument flight rules, IFR, the pilot must have obtained a clearance, must be in contact with appropriate controllers, and must adhere to the clearance unless a change is agreed to by the controller. All IFR airplanes are separated from each other, but they are not separated from VFR aircraft by air traffic controllers. Any separation from VFR traffic is by regulation: if the weather is bad, the VFR flying would be prohibited; if operating above 18,000 feet, VFR flying is prohibited; if flying in a TCA or an Airport Radar Service Area, VFR traffic would be controlled. Controllers give IFR aircraft information of VFR aircraft to the maximum extent possible, but they can't be expected to see and call all traffic all the time, especially in busy areas. As an aid to controllers, a "conflict alert system" warns if two IFR aircraft (or an IFR aircraft and a VFR aircraft on which the controller has instituted computer tracking) are on conflicting courses. All air carrier airplanes, and most of the truly active general aviation airplanes, have transponders with altitude encoding that give the controllers information on location and altitude. All aircraft flying above 12,500 feet are required to have this equipment. When an aircraft is given a transponder code, then the specific airplane will be identified as such on the controller's scope.

Visual separation is used with IFR aircraft when they have

reported other traffic in sight, as in the collision between the PSA 727 and Cessna in chapter 2.

Air traffic control is handled from different places. The control tower is the runway referee. At busier airports there is a terminal radar control facility that uses radar to sequence arriving traffic (this function is called approach control) and departing traffic (departure control) for the tower or for the en route controller who works in an air route traffic control center. The center controllers also handle approaching and departing traffic at airports that do not have approach and departure control available from a terminal facility.

That's an outline of the framework of the air traffic control system. It's the best one in the world, with the highest capacity, and while perfect safety is always desirable, it's unlikely, though we should never quit striving for perfection. Collisions that have happened illustrate where work can or has been done.

### Likely Spot

A Beech 99 commuter airliner and a single-engine Rockwell Commander collided in clear weather. The Beech had just departed San Luis Obispo, California, en route to San Francisco, and was climbing on a westbound heading. The Commander, on a training flight, was descending toward San Luis Obispo on an eastbound heading. The airplanes collided head-on at an altitude of about 3,400 feet.

The airliner had filed an IFR flight plan but because of the clear weather the captain had opted to depart from the uncontrolled airport VFR, to contact the center controller once aloft

and then get an IFR clearance. Neither an airline employee nor
the other person operating a radio on the frequency that would
have been used to broadcast traffic information recalled hearing
any radio transmissions from the flight; however, the pilot of
an airplane that preceded the flight in departing said he heard
them say that they were "departing runway two-nine straight
out."

After takeoff, the crew of the airliner called Los Angeles Center but because of other communications the controller didn't
answer for 40 seconds, at which time he told the flight to "go
ahead." The crew reported that they were climbing through 2,700
feet, "IFR to San Francisco." The radar controller assigned the
flight a transponder code and a few seconds later the controller
reported to the flight that it was "in radar contact six northwest
of the San Luis Obispo airport, say altitude?" The flight answered.
"Three thousand one hundred, climbing."

Then the controller cleared the flight to San Francisco, as
filed, to climb and maintain 7,000 feet. The flight acknowledged
by reading the clearance back. According to the controller, he
lost radar contact with the flight a few seconds after the clearance
was read back. The controller tried to establish radio contact but
wasn't successful.

Radar data from a track analysis program indicated that the
collision occurred at 3,400 feet. The fuselage and engine of the
Commander cut off the cockpit roof and fuselage crown of
the airliner; the safety board concluded that although the collision
was virtually head-on, the Commander was in a slight descending
right turn at impact. The board found that, based on the physiological limitations of the human eye, under ideal conditions
the pilots of the airliner theoretically could have seen the Rockwell Commander as early as 17 seconds before the collision; based

on the same considerations, the pilots of the Rockwell Commander theoretically could have seen the airliner as early as 23 seconds before the collision. The total time required for a pilot to sight an object, recognize that it is a collision threat, start an evasive maneuver, and have the airplane respond has been estimated to be about 12.5 seconds.

Two air traffic controllers involved said that there was no radar return from the Rockwell Commander on their scope. However, the safety board said that the evidence indicated that it was on the scope. There was a traffic situation demanding of attention in another area of the control sector; as a result, controller attention was concentrated on that part of the scope. Also, the airline flight wasn't identified on the controller's scope until 28 seconds before the accident.

The pilot of the Commander did not inform the center that he was going to be flying toward the airport. However, he did call on the appropriate frequency for San Luis Obispo and announce that he was over a radio fix six miles from the airport, inbound. However, the board found that the airliner had stopped monitoring this frequency when five miles away from the airport.

The NTSB faulted both pilots for not following the recommended communications and traffic advisory practices for uncontrolled airports. The physiological limitations of human vision and reaction time, as well as the short time available to the controller to detect and appraise radar data and issue a safety advisory, were cited in the probable cause along with an airline policy that required its pilots to tune one radio to the company frequency at all times. (With one on the air traffic control center frequency and one on the company's, they apparently had none on the traffic advisory frequency.)

## Big Sky

The sky is a big place, and on a clear day you can see a great distance, but this accident illustrates how two aircraft can be headed for the same little piece of airspace and a combination of relatively minor events can prevent the pilots from realizing this and taking the ever-so-slight evasive action necessary to let the two aircraft pass safely in the sky. You don't have to miss by much, just a little, but if you hit, it can all end quickly.

## Big v. Small

As was seen in the San Diego collision between a Boeing 727 and a Cessna 172, and in the 1960s' collisions between a twin Cessna 310 and a 727, a twin Beech Baron and a DC-9, and between a Piper Cherokee and a DC-9, where all airplanes were lost, a large airplane doesn't automatically have a significant advantage in a midair collision. In fact, relatively large airliners have in the past crashed after colliding with birds. But there have been cases of a large airplane hitting a small one and surviving. One that occurred a number of years ago, in 1971, prompted a look at weaknesses in the system, some of which have since been corrected.

The Boeing 707 had operated from San Francisco to the vicinity of Linden, New Jersey, without incident. It was cleared to descend to 3,000 feet, was being vectored for an approach to Newark, and had been assigned a heading of 180 degrees to assure adequate spacing behind preceding IFR traffic. The approach controller then said, "American 30, traffic at 12 o'clock less than a mile, northeastbound slow." (Slow simply means that the speed of the aircraft is slow in relation to others.) The flight reported, "No contact." On the flight deck, the following conversation ensued:

Everything sure is murky up here.
Boy, it is, and I suppose it's VFR.
Well, another thousand feet down is, but I hope nobody . . .
[Sound of object striking airplane]

The flight then transmitted that they had been hit by the other airplane. After the collision, the crew executed a series of shallow turns to determine the response of the airplane to flight controls and to assess damage. About 17 minutes after the collision, the airliner landed safely at Newark. The Cessna crashed, killing both occupants.

The flightcrew stated that they were all scanning ahead for the reported traffic when the head-on silhouette of a small airplane became suddenly visible through the haze. The airplane collided with the left wing of the airliner before evasive action could be initiated. The small airplane was a two-place Cessna 150, operating on a training flight.

This collision occurred in an area designated by the flight school as a training area. The school had advised students and instructors to remain below 3,000 feet in this area and, indeed, the collision occurred at 2,700 feet. However, no regulatory authority had been exercised by the FAA in the establishment of this training area and the air traffic controllers working the sector that encompassed the training area had no official documents or charts outlining the training area.

The reported weather at Newark was good, with broken clouds at 3,300 feet, an overcast at 8,000 feet, and surface visibility of eight miles. However, pilots operating in the area reported that the flight visibility beneath the coulds ranged from three miles down to less than one-quarter of a mile and varied horizontally as well as vertically.

While the 707 was cleared to fly at 3,000 feet, it was operating at 2,700 feet. The controller had no way of knowing this because the airplane was not equipped with an altitude reporting transponder. (It was not required to at the time.) The aircraft's altitude had varied between 2,800 feet and 2,700 feet in the time immediately preceding the collision. That deviation below the assigned altitude and variation has to be considered in context, because if the crew was devoting a lot of outside attention to looking for the other aircraft, as they should have been, they would have been paying less attention to flight instruments and the altitude variation could have been a result of this.

In the Cessna 150, the student was likely undergoing training in instrument flying because he was apparently wearing a hood, to restrict his vision to the instrument panel, at the time of the collision. So only the flight instructor would have been looking out for other traffic.

As a result of this accident, the NTSB recommended to the FAA that procedures be established for the reporting and coordination of practice areas used by all civilian flying schools for flight training. The FAA rejected this recommendation but since that time the coordination between flight schools and air traffic control facilities has become much more active and, in general, the conflict between aircraft on training flights and airline aircraft operating to and from major airports has been minimized.

### SEE-AND-AVOID

In its analysis of this accident, the NTSB said, "The weakness of the see-and-avoid concept of collision avoidance has been illustrated once again by this accident."

Most general aviation spokesmen vehemently defend see-and-avoid as a viable means of aircraft separation whereas airline pilots and especially airline management aren't so sure. It is often suggested that the airline interests would just like to get rid of all the small airplanes so they can have exclusive use of the airspace, but it is doubtful that many really have that motivation. It is a safety issue that must be continually evaluated, and the electronic collision avoidance system will be an aid when and if it comes into common use.

## Military Aircraft

In the past there was a record of military airplanes colliding with civilian aircraft, both general aviation and airline. This has been addressed with the institution of a lot of restricted airspace for military operations as well as with the designation of military operating areas—not restricted, but warning in nature. All military aircraft are supposed to operate IFR except when in a designated area or on one of the low-level practice routes that are shown on charts. This has apparently helped reduce the conflict between high-speed military aircraft and civilian airplanes.

A Marine F-4 (Phantom) jet and an airline DC-9 collided in California in 1971, adding impetus to many of the changes that were made to minimize conflict. The DC-9 was under radar control and was climbing to Flight Level 330 (33,000 feet) after departing from Los Angeles. The F-4 was being flown at approximately 15,500 feet under visual flight rules. The weather was good. The last communication from the DC-9 was routine. The F-4 had been plagued by trouble throughout the cross-country operation that it was on at the time of the collision. The transponder of the aircraft was inoperative and there was a problem with the oxygen system, meaning that it would have to operate at a relatively low altitude where the fuel burn would be higher, thus necessitating more fuel stops. After a fuel stop at Naval Auxiliary Air Station Fallon in Nevada, the pilot proceeded toward Marine Corps Air Station El Toro in California. Repairs to the aircraft had been unavailable at Fallon and the squadron duty officer had advised the pilot to fly on to El Toro at low altitude.

The flight initially climbed to 7,500 feet, and then to 15,500 feet to clear some mountains along the way. After the mountains, the flight descended back to 5,500 feet and remained at that altitude until near Bakersfield. A position report was made, and the crew checked the weather at El Toro. Then they decided to make an alteration in plan and fly farther east, over Palmdale, to avoid anticipated heavy traffic over Los Angeles.

The flight continued at low altitude, about 1,000 feet above the ground, until about 15 miles northwest of Palmdale. Then, because of deteriorating low-level visibility, the F-4 climbed back to 15,500 feet, which took less than two minutes at maximum power. Right after levelling off, the pilot executed an aileron roll maneuver and then settled into level flight. There were but about 30 miles to go to El Toro. The radar intercept officer in the back seat was operating his radar in the mapping mode, but because the air-to-air detection capability on the unit was degraded, no airborne targets were seen on the radar.

Just before the collision, the radar intercept officer observed the DC-9 in his peripheral vision approximately 50 degrees to the right and slightly beneath the F-4. He shouted to the pilot who had apparently already seen the DC-9 and started an evasive roll. He did not see the DC-9 take any evasive action.

The two airplanes collided, and after the collision the F-4 began to tumble violently. The radar intercept officer waited about five seconds, and, after seeing numerous warning lights in the cockpit, he ejected from the aircraft. The exit was successful and he parachuted to the ground without injury. The pilot in the front seat was unable to eject, and he was killed, along with the passangers and crew of the DC-9 which crashed out of control shortly after the collision.

The air traffic controller didn't give the crew of the DC-9 a traffic advisory on the F-4 because he did not see the aircraft on the radar screen. Because the transponder wasn't working, it would have been difficult to see. Later test flights showed that tracking continuity was poor and that the primary target of an F-4 flying that route at that altitude was visible on radar less than 50 percent of the time. It was demonstrated that the primary target alone was not of sufficient strength to assure notice by a controller who was not aware that the airplane was in the area.

The NTSB found this accident to be "another example of a heterogeneous mix of VFR and IFR traffic, with each air-craft complying with applicable regulations, resulting in a midair collision."

The probable cause of the accident cited ". . . the failure of both crews to see and avoid each other but recognizes that they had only marginal capability to detect, assess, and avoid the collision. Other causal factors include a very high closure rate, comingling of IFR and VFR traffic in an area where the limitation of the ATC system precludes effective separation of such traffic, and failure of the crew of the BuNo458 (the F-4) to request radar advisory service, particularly considering the fact that they had an inoperable transponder."

There were a flurry of recommendations following this collision, and many were followed, meaning that the chance of a midair between a military airplane and an airline aircraft has been lessened considerably.

## Near-Collisions

For every actual collision, there are a lot of near-collisions, and these always get good play in the press. (They are sometimes called a "near-miss," which is not really a proper description. If they nearly missed, that would mean they hit, wouldn't it?)

Two near-collisions that got everyone's attention involved Vice President George Bush flying aboard Air Force Two. (Air Force One and Two are designations given to aircraft when the president or vice president are aboard.)

A Cessna 310 operating on an IFR flight plan from Green Bay, Wisconsin, to Annapolis, Maryland, was at an assigned altitude of 13,000 feet. Air Force Two was also on an IFR clearance and was traveling from Cleveland to Washington. Air Force Two had been assigned an altitude of 8,000 feet and when control of the flight was transferred to the Cleveland air route traffic control center, the aircraft was cleared to climb to Flight Level 230 (23,000 feet).

Both airplanes had been identified and were being observed on radar. Both were southeastbound with Air Force Two behind but overtaking the Cessna. The big jet's crew had been climbing it at 250 knots indicated airspeed when below 10,000 feet, in compliance with the applicable rules, and the rate of climb there had been 1,500 feet per minute. When Air Force Two passed through 10,000, the speed limit no longer applied and the aircraft was accelerated to 320 knots while the rate of climb increased to 3,300 feet per minute.

As Air Force Two continued its climb, the traffic control computer's conflict alert function activated, alerting the controller to a possible conflict. The controller instructed Air Force Two to maintain 12,000 feet. The crew acknowledged and said they were passing 12,200 feet when the instruction was issued and that they would descend back to 12,000 feet.

Recorded radar data from the air traffic control center indicated that Air Force Two reached an altitude of 13,000 feet before descending back to 12,000 feet and that a minimum slant range distance of .25 nautical miles existed between the two airplanes at one point. At this time, Air Force Two was 600 feet below the Cessna, descending and passing to the right.

In its investigation, the safety board determined that an operational error occurred because of the unsatisfactory performance of the individual air traffic controller. The board said that the controller used poor judgment and poor control technique when he cleared Air Force Two to climb through the altitude being maintained by the Cessna. No mention was made of the significant change in rate of climb and airspeed on the part of Air Force Two after it passed through 10,000 feet. Though normal, and what would have been expected, it was possible that the controller could have based the clearance for Air Force Two's climb on its speed and rate of ascent as observed below 10,000 feet.

That violation of the minimum prescribed ATC separation of 1,000 feet vertically and five miles laterally occurred with both aircraft flying in instrument meteorological conditions (jargon for "clouds") so the only chance at separation was through air traffic control. System errors like this occur on a fairly regular basis, but the separation standards are such that there is some margin for error—enough that an en route collision between aircraft on IFR flight plans is almost impossible to find in the record book.

**NEXT**

Three weeks later, Air Force Two was involved in a near-collision incident of another kind.

The sky was clear and the visibility excellent as the jet was approaching Boeing Field in Seattle. A controller had instructed the crew to descend to 3,000 feet and subsequently cleared it for the approach, to cross a navigational fix at or above 2,600 feet. Then the controller advised Air Force Two of traffic at its 10 o'clock position, two miles distant. The flight acknowledged but said that the traffic was not in sight. A second advisory was issued on traffic at 11 o'clock, a mile and a half. Again the crew acknowledged but said that the traffic was not in sight. Shortly afterwards, Air Force Two advised the controller that they were changing to the Boeing Tower frequency. The approach controller called the tower controller on a landline and advised that Air Force Two would be calling and that they didn't have the traffic in sight that was off to the left. When Air Force Two reported in to the tower, they advised that they had just had to take evasive action to avoid traffic.

Air Force Two's aircraft commander, who was seated in the right (copilot's) seat, first observed the other airplane, a single-engine Mooney, out of the cockpit's left side window. The Mooney appeared to be level with his aircraft and on a collision course. He estimated it to be from 1,000 to 1,500 feet away when first sighted. The commander assumed control of Air Force Two, retarded the power, and pushed the nose down slightly. The Mooney passed directly over the midsection of Air Force Two about 100 or 200 feet above the airplane.

The pilot of the Mooney said that he was not aware of the near-collision until the FAA contacted him about two weeks after the occurrence. Because of problems with his radio equipment, the pilot said that he was not able (nor was he required) to contact air traffic control for traffic advisories.

In this case everything worked almost as it should. The controller spotted the traffic and told the crew of Air Force Two about it. But the crew didn't see the airplane soon enough to avoid it without taking what they deemed "evasive" action.

For the Mooney pilot, the board's investigation determined that the pilot "used poor judgment in initiating a flight in close proximity to the Seattle TCA with both radios inoperative." That was a bit of a strange finding because the pilot did not violate the TCA airspace and most pilots wouldn't consider that they were using poor judgment when they were abiding by all the rules. Theoretically, at least, restricted airspace is just that and the measure of success is avoiding it—even if by a scant margin.

### Subjective

Near-collisions are a very subjective thing. One pilot's near-collision might be the other pilot's friendly passing in the sky. How such an event is viewed is probably related more to the startle factor than anything else. If you see another airplane in plenty of time and go close behind, or over or under, the passing might seem routine. But if the other pilot does not see you until the separation between airplanes is minimal, that pilot might feel it was a near thing and might take abrupt evasive action at first sighting. There is on record at least one collision between two airplanes that would not have collided had not one of the pilots taken evasive action. Because of a combination of visual illusions, the pilot thought the airplanes were on a collision course even though they were not.

When flying at higher altitudes, where vertical separation of 2,000 feet is provided between all traffic, it's honestly difficult to make a quick decision on whether an airplane ahead is higher or lower. When flying around at low altitudes, it's easier to judge the relative height of the other airplane, but harder to see it against the ground if it's a bit lower. Most pilots avail themselves of radar traffic advisories, which help, but they are no guarantee of separation from other aircraft.

## VFR v. VFR

The collisions we've looked at so far involved one aircraft flying VFR and one flying IFR. The great majority of the collisions involve two airplanes flying VFR, most are general aviation aircraft, and most occur near an airport. There are, however, en route VFR v. VFR collisions, and one in California involving a commuter airline Twin Otter and a Cessna 150 illustrates how getting traffic advisories when on a VFR flight doesn't insure anything.

The Twin Otter was flying from Ontario, California, to Los Angeles International Airport on a VFR flight plan. Along the way, the Twin Otter crew called the Los Angeles arrival radar controller and reported over Rose Hills, a local landmark. The flight was in radar contact 23 miles east of the airport and was cleared into the Los Angeles terminal control area for an arrival on Runway 24 left.

A bit later the arrival controller verified that the Twin Otter was leaving 2,600 feet and advised the crew that they had traffic about 5.5 miles ahead, climbing from 1,500 to 3,000 feet. The traffic was a police helicopter and the controller told the flight that he would point out the helicopter again when it was closer. The controller also asked the flight to report when they had the helicopter in sight. They said they would, and this was the last known transmission from the Twin Otter. The controller transmitted subsequent advisories about the helicopter, but the flight did not reply, and its target disappeared from the radar scope.

A Cessna 150 based at Long Beach, California, had departed from that airport about 20 minutes earlier on a training flight with an instructor and student on board. There were no radio contacts with this aircraft after it finished talking with Long Beach Tower, nor were any required.

According to information available from the Los Angeles radar, the Twin Otter was descending about 300 feet per minute on a magnetic course of 250 degrees with a groundspeed of about 150 knots. From the damage to both aircraft it was determined that the Cessna was on a northerly track at the time of the collision. Flight tests showed that an aircraft which followed the presumed track of the Cessna 150 in the area of the collision did not produce a primary radar return (the 150 had no transponder) on the controller's display. The controller stated that he did not see any primary (nontransponder) returns on his display.

The weather was excellent, so see-and-avoid should have worked. However, the probable closure angle of the two airplanes placed the Twin Otter 57 degrees to the right of the Cessna.

According to the NTSB report, a pilot's scan for traffic normally will range about 45 degrees to either side, and the Twin Otter would also have been masked by the Cessna's right wing; therefore, its pilots would not have been looking in that direction unless they had been alerted to the presence of traffic.

The ability of the Twin Otter crew to see and avoid the Cessna was affected by two factors. The sun was setting, and the Cessna was between the Twin Otter and the setting sun, making it difficult if not impossible to see the Cessna. Second, the word from the controller that there was a helicopter ahead, climbing through the altitudes through which they were descending, would just naturally have drawn the Twin Otter crew's attention to the area ahead of the aircraft, where the helicopter was reported.

The statement of one witness indicated that neither aircraft made any abrupt evasive maneuver prior to the collision; therefore the safety board concluded that neither crew saw the other aircraft prior to impact, or in sufficient time to attempt an evasive maneuver.

### FIXATION

It is very difficult not to fixate on an area where traffic has been called. You *know* there is something out there and the challenge is to find it. The military and the airlines have scan training programs, and this is stressed in general aviation. It is a necessary item in the scheme, but there will always be times when the ability to see and avoid other traffic is compromised by other factors, such as the visibility limitations of the aircraft or the setting sun.

## Better Ending

Not all midair collisions have an unhappy ending, nor do all happen on clear days. Again in 1971 (that was a bad year), the collision of a Boeing 707 and a Cessna 150 at night, near Los Angeles, was an example of an unusual accident.

The 707 was operating from Hawaii to Los Angeles and established communication with the Los Angeles arrival controller while descending to 8,000 feet on a heading of 360 degrees. The flight was cleared for further descent, and there was discussion about which runway the flight would be using and the traffic that it would be following. The controller said, "Do you see the traffic you're following at 11 o'clock and five miles?" The captain replied that they had seen the aircraft, but did not see him at the present time. The flight was then advised to adjust airspeed to 180 knots.

Next, the flight was advised of traffic at 11 o'clock, one mile, northwestbound. The captain simply replied, "Roger." The controller asked the flight if they could adjust their pattern and shortly after this the captain reported that he believed they had just had a midair collision and that they were declaring an emergency. The flight was cleared direct to the airport and to land, which it did successfully. After landing, the captain stated to the ground controller, "I believe we got a small airplane out there just as we were turning final, looked like a Cessna."

It was indeed a Cessna, one that was on a night training flight. It was the student's first familiarization flight and the aircraft was near Compton Airport, its point of departure, and between 3,000 and 4,000 feet as the instructor told the student to turn to a southeasterly heading. The collision occurred just as the aircraft reached the southeasterly heading. The instructor then took over the controls and landed the severely damaged aircraft in a field adjacent to a lighted all-night golf course.

The NTSB reported that the instructor in the Cessna said the aircraft had all its position lights and rotating beacon on at the time of the accident. He did not see the jet before the collision. Only partial control of the aircraft was available and the engine ceased functioning after the collision. The instructor said that he believed his right ankle was broken in the collision but he could not remember when his left ankle was broken. The student pilot said that he saw a bright light in the lower front part of the right door window a split second before the collision.

At the time of the collision, the airliner and Cessna were on headings of 340 degrees and 135 degrees respectively. That's almost head on. The impact damage showed that the main wheels of the Cessna contacted the leading edge and upper surface of the right wing of the airliner. Although the Cessna had its lights on, the crew of the descending Boeing was looking for the aircraft against the lights of the Los Angeles area, making sighting difficult. Also, the airline crew was looking for traffic it was to follow and the board believed that the last transmission about traffic to follow may have distracted the crew from scanning for the Cessna just prior to the collision.

The NTSB determined that the probable cause of the accident was the minimum opportunity for the flightcrews to see and avoid the other aircraft due to the background lights behind the Cessna and the decrease in the Cessna's pilots' visual field resulting from the aircraft's wing while turning. In a companion recommendation, the NTSB said it believed it an unsafe practice to engage in student familiarization flights in high-density traffic areas at night.

## Mistaken Identity

Another collision of the '70s shows how the mix between very
high-speed aircraft and slower ones can result in problems in
middle-level airspace. At 10,000 feet above sea level, the 250-
knot speed limit goes off and positive control airspace doesn't
come into effect until 18,000 feet, so the speed mix in these
altitudes can be great. It must be added that there is not a lot of
VFR traffic in this airspace, and there is now a requirement for
altitude reporting transponders above 12,500 feet, so at least above
that altitude the controllers have altitude information on VFR
traffic.

A twin turboprop Commander was flying from Phoenix, Ar-
izona, to Butte, Montana. The pilot was flying VFR, but when
he was in the vicinity of Cedar City, Utah, he called a flight
service station and filed an IFR flight plan. The aircraft was then
told to contact Los Angeles Center for further clearance.

The pilot called the center and was asked to "Ident, say altitude." The pilot replied, "Okay, we're squawking 1400." (The
ident function of the transponder changes the appearance of the
return on the transponder's scope, for positive identification. At
the time of this accident, 1400 was the code used by VFR aircraft
operating above 10,000 feet.) The pilot added that he was at
17,500 feet and would like to fly at 18. The controller told the
pilot that he didn't see him and asked for a position in relation
to Bryce Canyon. The pilot did not reply.

An Air Force KC-135 tanker from Grand Forks Air Force
Base in North Dakota had arrived at an airborne refuelling point
a bit earlier, and had been cleared into an altitude block of FL
180 to 210 (meaning the aircraft could fly at any altitude within

that block), and had asked for a delay in the area to await two
F-111s. The altitude proposed for the refuelling operation was
FL 200, and the tanker advised one of the F-111s of its position.
It was dark and the tanker was illuminated with wingtip and
tail position lights, anticollision beacons on the top and underside
of the fuselage, engine nacelle floodlights, underbelly lights, refuelling boom lights, and a boom nozzle light.
The two F-111s joined in formation and then planned to
rendezvous with the tanker. Because of a problem with the navigational equipment in one of the F-111s, lead of the two-ship
flight was changed and the aircraft were 15 to 17 miles off the
published refuelling track as the maneuvering started.
The pilot of the lead F-111 later stated that he saw a light
and beacon and that, "As far as I was concerned (they) were in
the same position." He called a tallyho (in sight) at 12 o'clock
and "followed that beacon the rest of the way."
The weapons system officer of the lead F-111 stated that he
got a radar lock on on the aircraft at about five miles and noticed
that they had started a turn to the right to chase the light that
they saw in front of them. The pilot of the other F-111 saw a
beacon ahead when the lead pilot called a tallyho. He thought
it was the tanker; however, after that time he just concentrated
on flying formation with the other F-111.
As the two F-111s continued to close on the target, the weapons system officer told the pilot that the range was two miles.
The pilot in turn requested that the tanker increase his speed to
the refuelling airspeed of 305 knots indicated airspeed. The WSO
called the target at 4,000 feet radar range and said, "It looks like
we have a fast overtake." The pilot rechecked his instruments,
confirmed that he was at Flight Level 180, and noted that his
airspeed was at 320 knots and decreasing. The WSO called a
range of 2,000 feet and indicated that they were closing fast. The

pilot said that he again checked airspeed and altitude. He then looked up and saw a white light and what he thought was the right outboard engine pod of a KC-135. He said he pulled back on the control stick and collided with what he thought was the tanker. They had actually collided with the turboprop after overtaking it from the rear at a closure speed that was calculated by the NTSB to be 158 knots.

Both aircraft were destroyed in the collision and fire that followed, but the F-111 crew was able to use their escape capsule and parachute to the ground. The other F-111 was not involved in the collision, which occurred at 17,900 feet.

The tanker was at Flight Level 222 on a northwesterly heading; the Commander was on a northerly heading, apparently at 17,900 feet, just below the floor of positive control airspace, awaiting an IFR clearance. The collision occurred outside the refuelling track assigned to the military aircraft.

In its report, the safety board explored the hurried atmosphere that followed the change in lead of the F-111 flight. There was also an undetected position error, caused by a problem with the original lead aircraft's navigational gear, that placed the aircraft out of the refuelling track. Then the F-111 lead pilot misidentified the lights of the Commander as those of the tanker.

With the tanker at FL 200 and the F-111s at FL 180, the faster aircraft coming in to be refuelled would be making their approach 2,000 feet below the tanker, to pull up when close and when positive identification was made. With the tanker 2,000 feet above, it would have been in a relative position 45 degrees above the F-111s, or high in the windshield, when the overtaking aircraft was at a range of 2,000 feet.

The board said, "The implications of this visual phenomenon are well known to pilots who are experienced in air refuelling rendezvous. Because the beacon remained in the center of his windscreen and the aircraft had not climbed from FL 180, the pilot . . . should have recognized that a collision was imminent and that the beacon he was approaching was not Toft 51 (the call sign of the tanker)."

The safety board wasn't able to determine why the Commander was at 17,900 feet. The proper VFR altitude for the aircraft would have been 16,500 feet.

The probable cause of the accident was deemed by the NTSB to be the F-111 pilot's "misidentification of the Turbo Commander as a refuelling tanker with which he intended to rendezvous. Contributing to the misidentification was his failure to use prescribed procedures and techniques during rendezvous with a tanker aircraft for refuelling."

## Ground Collisions

The most tragic collision in aviation history occurred on the ground, at Tenerife, killing 581 people. It involved two 747s, there because of a bomb scare at another airport. There was a misunderstanding and one aircraft attempted to take off before the other had cleared the foggy runway while taxiing. A similar collision, less publicized because there was far less loss of life, happened in the U.S. This one is also from the '70s.

The weather at O'Hare airport was foggy. The visibility was a quarter of a mile and controllers couldn't see aircraft on the taxiways and runways.

A CV-880 jet had landed after a flight from Tampa, Florida, and was asked to report when clear of the runway. The crew did, and was told to contact ground control.

The flight called: "Delta 954 is with you inside the bridge

and we gotta go to the box." (The box is a holding area on the airport, where incoming flights would await a gate assignment.) The controller replied, "Okay, if you can just pull over to (the) 32 pad." The first officer replied, "Okay, we'll do it." The controller made an entry on a scratch sheet which he later stated was to remind him that he had sent the CV-880 to the Runway 32 *right* pad to hold awaiting a gate assignment. The captain taxied the aircraft en route to the Runway 32 *left* runup pad, where both the captain and first officer thought the controller had cleared them to go.

In the meanwhile, the tower had cleared a DC-9 for takeoff on Runway 27 left. The captain of this aircraft reported to the tower that he was beginning his takeoff roll. The first officer was flying and everything was normal to the point where the captain called "rotate." At that moment, the captain saw another aircraft ahead on the runway and he immediately assisted the first officer in applying additional control pressure to gain altitude in an attempt to clear the other aircraft. The attempt was unsuccessful and a collision occurred with the CV-880 that was crossing the runway. After the collision, the DC-9 captain decided that his aircraft could not maintain flight so he took control and flew the aircraft back onto the runway. The DC-9 was destroyed in the crash and subsequent fire and 10 of its 41 passengers were killed. The CV-880 was damaged substantially.

How could something like this happen at a major airport, which should have the best equipment and procedures? As usual, there were a lot of factors that led the CV-880 to be on the runway at the time the DC-9 was taking off. The NTSB theorized that the sequence of events probably started when the CV-880 crew listened to the transcribed automatic terminal information service broadcast before beginning their approach to O'Hare. The broadcast announced that Runways 14 right and left were being used for departure. The crew wasn't told when the departure runway was changed to 27 left.

The safety board was of the opinion that once the CV-880 was down and talking to ground control, the controller did not hear the words "inside the bridge" in the transmission from the flight. Had he heard their position, he would not have directed the flight to the 32 right pad, his stated intention. That would have been against the flow of traffic. If the controller had intended to direct the aircraft to the 32 left pad, he would have had to coordinate with the local controller for the aircraft to cross 27 left, the active takeoff runway, and this was not done. The review of the recording of communications showed that the transmission from the flight was both audible and intelligible, so the board couldn't determine why the controller didn't hear these words. Communications in total were deemed not good, and in the opinion of the board, "There was a need for a request for additional information and for clarification on the part of both the flightcrew and the controller."

The tower was equipped with an airport surface detection equipment radar system, and controllers testified that during periods of low visibility this was used almost exclusively by the local controllers to determine whether approaching aircraft had landed or executed a missed approach, when and where landing aircraft were clear of the runway, and when departing aircraft began and completed the takeoff. The controllers also testified that they considered the ground radar equipment unreliable for the identification of airport traffic movements because of blind spots, the inability of the equipment to distinguish aircraft from other vehicles, and the derogation of target definition during periods of moderate to heavy precipitation. The ground controller on duty at the time of the accident was not required to be qualified, nor was he qualified, in the operation and use of the radar system.

He did use it in assisting another flight in locating the box, but not to identify the position of the CV-880.

The NTSB found the probable cause of this one to be "the failure of the traffic control system to insure separation of aircraft during a period of restricted visibility. This failure included the following: (1) the controller omitted a critical word which made his transmission to the flightcrew of the Delta CV-880 ambiguous; (2) the controller did not use all the available information to determine the location of the CV-880; and (3) the CV-880 flight-crew did not request clarification of the controller's communications.

## Bad Day at Atlanta

There have been several documented instances of control system errors that have led to less-than-prescribed separation in flight. One clear day in Atlanta saw a mix-up of such proportions that the NTSB did a special investigative report on the event.

First, a little primer on how traffic is handled at the Atlanta airport. When an aircraft approaches the Atlanta area, it is generally under the control of the Atlanta Center, which deals primarily with en route traffic. The en route controller then "hands off" the aircraft to a feeder controller in the Atlanta terminal radar control facility. This is done electronically, and the receiving controller must acknowledge the handoff by a keyboard action. The feeder controller's job is to sequence both VFR and IFR traffic into an orderly flow for handoff to a final controller. The final controller's job is to sequence and position aircraft for the final approach, turning aircraft over to the tower (or local) controller for landing clearance. The Atlanta north feeder controller works two corridors of inbound traffic, one from the north-west and one from the northeast. The corridors converge about seven miles north of the airport. The feeder and final controllers are located in the same room but direct voice communication between them is not possible. An intercom system must be used. They communicate with aircraft on separate frequencies so cannot hear the others' radio communications with pilots.

### LOTSA TRAFFIC

When this event started, four aircraft were nearing positions where they would be handed off to the final controller by the feeder controller. At about the same time, the final controller was handling five aircraft, three inbound air carriers from the north and two general aviation aircraft from the south. The two air carriers nearest the airport were sequenced and landed and the final controller then turned his attention toward the third air carrier, a DC-9, and the two general aviation aircraft. In the meanwhile, he had accepted two more inbounds from the north.

The final controller next told one general aviation aircraft, which was at the lowest altitude and nearest to the airport, to turn to the west and onto final approach, in front of the DC-9 that was descending from the north. The controller then realized that the DC-9 would overtake the slower airplane so he turned the DC-9 north. This occupied the controller's attention as an eastbound L-1011 and a southbound 727 were entering his airspace.

The controller accepted the L-1011, which was descending through 8,500 feet, even though radio communications with that aircraft hadn't been established. The 727 had descended to 7,000 feet and was on the final controller's frequency; however, since the final controller had not accepted the handoff, no positive action was taken by either the feeder or final controller to maintain separation between these two aircraft. The pilot of the L-1011 saw the 727 and increased his rate of descent to pass below it with about 400 feet of vertical and a quarter of a mile lateral separation. This activated the conflict alert system and the feeder controller was told to halt handoffs to the final controller, and the center was told to hold off all traffic until the situation was resolved.

Next, the controller turned the 727 westbound and the L-1011 southbound and then west onto the final approach. After its turn, though, the 727 was too close behind a slower general aviation aircraft so the controller turned it north and then east, with the aircraft assigned to 7,000 feet. This put the 727 back in the feeder controller's airspace and had the ultimate result of placing the 727 in conflict with a second L-1011. The conflict alert went off again.

Meanwhile, the DC-9 that had been taken out of the sequence earlier was being vectored for final approach ahead of the first L-1011. But after the turns they were in almost the same position, with the L-1011 1,500 feet above the DC-9. So the controller turned the L-1011 to the north and into a conflicting situation with three other aircraft, one of which was the 727. The conflict alert for the L-1011 and 727 went off first, followed by conflict alerts on the other two aircraft. All four were within a two-square-mile area; one was talking to the feeder controller and three to the final controller. One aircraft passed less than a mile in front of another with 200 to 300 feet of vertical separation; two others were within 100 to 200 feet of the same altitude and close enough for the pilot of one aircraft to exceed the engine temperature limits in a hasty climb to improve separation. Two additional aircraft became involved in reduced separation situations that activated yet another conflict alert, but they didn't come close to colliding. After the aircraft all passed each other, vectors were issued to restore order to the flow of approach traffic. In a six-minute period, a series of near-collisions involving five aircraft had occurred.

### PILOTS

That the pilots were unhappy about all this is an understatement. They basically had to separate themselves for a while in what was supposed to be an environment of positive control. And it takes little imagination to see what might have happened had all the aircraft been inside clouds instead of flying on a clear day.

The NTSB's findings in their special report focused on the handling of the traffic by the controllers, on training of the controllers, on procedural deficiencies, on coordination, and on the way conflict alerts are displayed on the radar system.

Of all those, I think the procedural part is the most important to the big picture. When there is a good procedure for handling traffic, things tend to go well. When the procedure is weak, things don't go so well. This is especially true when controllers are mixing aircraft of different speed—which is almost always the case.

When flying into my home airport at Trenton, New Jersey, from the west, the procedure is to fly over Harrisburg, Pennsylvania; Lancaster, Pennsylvania; a navigation fix called Bucks intersection; then on to Trenton. Bucks is what you might call a feeder fix to the Philadelphia terminal area, in which Trenton is located. All that is great.

This, though, is one track to be used by aircraft of all speeds, which adds to the workload of the controller. The procedure is to have the slower airplanes fly lower than the faster ones, and the procedure tends to sometimes get knotty out in the airspace controlled by New York Center, as they prepare the flow of traffic into the Philadelphia area while at the same time coordinating it with other traffic.

Some controllers have trouble melding the descent characteristics of slower airplanes and faster airplanes to the point where aircraft have to be held at an altitude while an airplane below descends, or one might have to be turned. That diverts the controller's attention from the big picture and makes his work harder. It's a tough equation—a piston-powered airplane descending out of 18,000 feet at 240 knots groundspeed and 1,000 feet a minute rate of descent being overtaken by a 480 groundspeed at 3,000 feet per minute takes a moment of interpretation. The procedure that could be used, and that is used for the fast traffic, is to tell the crew to descend to cross certain points at (or at or below) certain altitudes. But that procedure is not, for some mystical reason, often used in this area for the slower airplanes. The result is that the controller knows at what point the fast airplane will be at certain altitudes, but must monitor the slower airplane and determine where it will reach certain altitudes. And any time workload is increased, so is the likelihood of error.

## Total

Reading about all these collisions, near-collisions, and glitches in the air traffic control system might make any user of the system, whether a pilot or a passenger, cringe. But it shouldn't. People in the air traffic control business are like pilots in studying incidents and accidents to prevent any repeats. From virtually every event comes a better procedure or system. On balance, the air traffic control system in the U.S. is not only the most efficient in the world, it is also probably the safest. Sure, people make mistakes now and again, but they are not frequent. And when mistakes are made, the design of the system and the margins that are built into it keep virtually all events in the incident column as opposed to the accident column.

# 9

# Superstition, Luck, and Coincidence

NEEDLESS TO SAY, superstition, luck, and coincidence have no place in aircraft operations. The key to a good record is in understanding risks and managing them. Still, aviation has a few things that stand out, things that might be charged off to more than happenstance.

One is the business about whistling in the cockpit. For years, it seemed that every accident report that included the transcript of a cockpit voice recording included the notation "whistling." That became so widespread, and got so much publicity, that it came to be considered a bad omen. If another crewmember, or a passenger in a front seat, started whistling, there would come a severe reprimand from the captain. It's almost as if the act of whistling could cause a wreck. The debate about whether whistling is a sign of fear or of complacency is not likely to ever be settled, but at least whistling in the cockpit seems to be on the wane.

## Who Is Flying?

A coincidence that does continue relates to which pilot is flying the airplane when it crashes. Most U.S. airlines follow the policy of having the captain fly one leg and the first officer the next. They don't swap seats. The first officer has a duplicate set of instruments and controls, and for someone who is used to flying from the right seat, it works just as well as flying from the left seat. When the first officer is flying the airplane, then the captain, in the left seat, performs the copilot duties.

If you will think back to the airline accidents that have been explored in this book, the first officer was flying in most of them. There is such a strong pattern that it has to be given consideration. Many contend that this has just been happenstance, but that is not a satisfactory way to dispense with such a strong coincidence.

Why might most accidents happen on the legs that the first officer is flying? A NASA safety publication, "Callback," offered some thoughts on the subject.

One came from an airline policy statement: "The risk in the first officer flying is not that the first officers are less capable, but that captains are less efficient in the assisting role. Confronted with the ultimate responsibility, captains are likely to be monitoring the operation so closely . . . and to be so involved that their assisting role may be neglected. . . ."

Another comment was included in an FAA study: "When the first officer is flying, the captain often fails to execute normal first officer functions and duties. . . ."

From material at an international aviation conference: "The roles of the pilots during the approach and landing were discussed. . . . In this connection it had been discovered that co-ordination between the pilots, which was satisfactory when the captain landed the aircraft, was often less so when the copilot was controlling the aircraft and the captain undertook the duties normally those of the copilot."

When the senior pilot on the flight deck, the person ultimately responsible for the safe operation of the aircraft, assumes the role of the assistant, is safety derogated? From experience, one would say that should not be the case.

Often, when flying in the right seat with another pilot flying my airplane, I feel that it's actually possible to have a better grasp of the big picture. Freed of the necessity to interpret the instruments on a second-to-second basis and make the required control inputs, I feel that the deal is better. The beginning of a possibly difficult situation is less likely to get by without notice. But there is a strong possibility that this is a false comfort. In looking at the airline accidents in which the first officer was flying, and that includes the more spectacular wind shear accidents, you have to wonder if the captain would have elected to fly through that patch of weather had his hands been the ones on the handlebars. Is there a chance that a captain would feel enough more in command, and enough more in touch with everything, that he would opt to continue when otherwise a diversion might be the choice? There is no way to know for sure, but I do know that at times I like it better with someone else flying in turbulent air, leaving me with plenty of time to study and analyze the information on the airborne weather radar and Stormscope (lightning detection equipment). Whether this contributes to safety, or leads to flight through an area that might otherwise be culled, is open to question.

### Study

The NASA group that produces the safety publication mentioned runs an aviation safety reporting system. Pilots and air traffic controllers report glitches to this group and if you make a little error unintentionally, and report it promptly to NASA, then any

FAA prosecution as a result of the error is blunted. The purpose of that, and it works, is to encourage the reporting of anomalies so a compilation can be made and used for educational purposes.

In a study relating to the exchange of pilot duties, 245 reports were examined. It was found in this base of information that captains were far more effective in detecting anomalies than were first officers, regardless of who was physically flying the airplane. Captains caught 33 percent of the foul-ups when they were flying and 35 percent when the first officer was flying. The first officer detected but 10 percent of the problems when he was flying and 15 percent when the captain was flying.

The primary error in this group of reports was what is called a "busted" altitude. That is, the flight was assigned an altitude and the pilot flying the aircraft failed to level off at that altitude. This appeared as a greater problem when the first officer was flying, which would suggest that the captain wasn't doing a proper job of monitoring altitude and making the required callouts to the pilot flying at specified times, before reaching an assigned altitude. But the captain was pretty good at catching the error after the fact.

The second-largest reported "error" resulted in a near-collision with another aircraft. In this type of event, the pilot not handling the controls would be the primary lookout and more of these occurred when the captain was flying the aircraft. Near-collisions were more a problem during departures with the captain flying, and during approaches with the first officer flying. The relatively small number of aircraft-handling anomalies occurred during approach and landing with the first officer flying.

In many ways a departure is less critical than an arrival. The airplane is flying away into a great big sky and small errors in aircraft handling are not likely to be critical. But on an approach, the aircraft is flying into the small end of the funnel and must reach the ground at a proper speed at a rather precisely defined point. This requires both more concentration inside the cockpit on the part of the pilot flying, and more concentration on monitoring the operation by the captain when the first officer is flying.

In its summary of this study, NASA's "Callback" publication stated that "when the captain was flying and the first officer was responsible for monitoring and radio communications, a greater number of near-collisions, takeoff anomalies, and crossing altitude deviations occurred. When the first officer was flying and the captain was responsible for the pilot-not-flying duties, a greater number of altitude deviations, near-collisions during approach, and landing incidents were reported. . . . The flightcrew operates more efficiently when the captain is flying than when he is performing not-flying duties. . . . The captains appeared to have a higher level of operational awareness than did the first officers . . . pilot handling skills were a minimal factor . . . the importance of the monitoring function was not well understood by either pilot. . . ."

## No Conclusion

That leaves us where we started. This is something that airline safety and training departments have to work on to minimize what is an obvious increase in risk when the first officer is flying. And while it might seem apparent that a cure could be in a decree that the captain does the flying and the first officer does the assisting, that would in the long term work against safety. Someday the older person in the left seat is going to retire and the pilot in the right seat is going to move over into the left and become captain. The experience gained in those years of role-swapping is essential to the upgrading process.

There is one other factor that I've often considered in relation to this subject, one that I've not seen mentioned in any study. There is no requirement that an airline captain be a flight instructor or have any experience teaching others how to fly. Yet in the role-swapping process, the captain is in effect becoming a flight instructor. This is especially true if the first officer is relatively inexperienced. Any pilot who has done a stint as a flight instructor develops sort of a second sense that is required for survival when teaching others to fly, and letting them go as far as possible with mistakes for the educational value of letting them see the results of the error in their ways. That instructors develop this sense is reflected in the relatively low accident rate in instructional flying. An instructor also learns that there are times when it is better to tell the student that it's time for him to watch instead of fly for a minute.

### Hazardous Air?

The Bermuda Triangle is one of aviation's leading superstitions and there are other areas where a mystic quality has been added to the fact that an above-normal number of accidents have occurred. These things are caused not by the supernatural but by what pilots try to do with airplanes in the area. In western Arkansas, for example, where the airliner crashed when trying to go around thunderstorms, it is a combination of area weather characteristics and mountains where many people don't think there are mountains. If an airplane disappears while flying over the ocean, it's natural that some mystery will be assigned to its disappearance because the wreckage of the airplane is unlikely to be found and the cause of the accident will almost certainly remain "undetermined." When they crash on land and the remains of the airplane can be subjected to close scrutiny, the probable cause can almost always be determined.


## *Survival*


Another superstition, or at least widespread feeling, is related to survival in an aircraft accident. Some feel that if you sit in the rear of the airplane, chances are better. That's usually based on the results of one accident, such as the L-1011 at Dallas/Fort Worth, where survivors were seated in the rear. But then comes an event like the British Airtours 737 at Manchester in 1985, where the aircraft stopped on fire and the survivors were from the front of the airplane. Experts can proclaim that one place or another in an airplane is safer, but in truth you'd have to ask the captain what kind of accident the airplane will be involved in to get a true measure of the safest place to sit. And then there are those who say that if you are in an airplane wreck, you've had it. Period.

There is no question that airplane crashes are more likely to be fatal than automobile accidents. In the latter, a small percentage of the accidents result in fatalities. In general aviation flying, almost 20 percent of the events classified as accidents result in fatalities, while a much smaller percentage, about 10 percent, result in serious injuries. The rest cause minor or no injuries. So, if you are in a general aviation accident, chances are 90 percent that you'll either die or not be hurt badly. Adding the third dimension that is flight simply means that the hit will either be hard, or won't amount to much. There's not a lot in between.

**AIRLINES**

In airline accidents, there tend to be few or no survivors any time the airplane winds up somewhere other than at an airport. The only modern-day airliner that I know of with a long history of off-airport crashes with survivors is the Fokker F.27 (also built as Fairchild FH-227), a high-wing turboprop twin with about 50 seats. These airplanes have been left impaled in houses, on mountain sides, and in other random areas, usually with survivors. Maybe it's the fact that the airplane has a relatively low landing speed, or maybe the high-wing configuration isn't as prone to spill fuel in a manner that causes a fire after the crash. Whatever, the F.27 has always stood out.

## On-Airport

In on-airport accidents, the passenger's protection is rooted in an FAA requirement that operators of aircraft with more than 44 seats must demonstrate that the aircraft can be totally evacuated in 90 seconds. This sounds good, and it is demonstrated, but in actual practice the evacuation of aircraft has often failed to realize the potential. Panic, alcohol consumption, passengers scurrying for carry-on baggage, the failure of furnishings within the aircraft, smoke, and fire can all hinder the evacuation of aircraft. In a special study on airline passenger safety education, the National Transportation Safety Board outlined some cases where evacuation was a problem.

A DC-10 ran off the end of the runway following a rejected takeoff. In stopping, the aircraft struck some buildings, the right wing and engine separated, and the tail section was engulfed in flames. The aircraft was at or near capacity, with 393 on board. All of the fatalities were in the aft cabin of the aircraft, where all survived the effects of the crash but many perished in the fire. According to the NTSB, the right aisle was clear and could have been used by these passengers to move to forward exits which were in use and clear. But for some reason they didn't use this aisle.

Many of the survivors of this accident indicated that their evacuation was not influenced by the safety information that had been presented. One said, when asked if the emergency instructions were of any value, "This information was not retained in a moment of crisis."

### Short at Night

An airline Boeing 707 landed 4,000 feet short of the runway at night and struck rocks, trees, jungle vegetation, and a lava wall before stopping. Fire broke out in the last 300 feet of the slide-crash process. But one of the 101 people on the aircraft was killed by the impact; all but four died in the fire that ensued. Survivors reported that before the airplane stopped, passengers rushed to exits in the front and rear of the cabin and they reported that they heard no evacuation instructions after the accident. In this case, the NTSB found that three factors adversely affected survival: (1) the flight attendants did not open the primary emergency exits; (2) the passengers' reaction to the crash; and (3) the attentiveness of the passengers to the pretakeoff briefing and to the safety cards.

## DC-9 on the Airport

A DC-9 crashed while attempting to go around after a balked landing; the aircraft hit the ground hard and slid 2,000 feet before stopping. The tail was broken off but the fuselage remained intact. Most of the aircraft seats failed and while nobody was killed, only 20 of the 107 people on board escaped with no injuries.

Passengers and crew reported that the seat failures, which occurred at impact, caused passengers to be thrown into adjacent seats, or pinned them between seats, between the floor and seats, or between seats and sidewalls. Failed seats were thrown into the aisles and against other seats. Overhead storage racks failed and dumped their contents; baggage and garments were in the aisles as passengers evacuated, and some stopped to retrieve possessions before leaving the aircraft. As many as 12 were still trapped inside when the fire department arrived. Clearly there would have been at least 12 fatalities had fire erupted after this accident.

## Sea Gulls v. a DC-10

A DC-10 carrying 128 passengers, all employees of the airline and all but one of whom had had at least some emergency training or familiarization with the aircraft, aborted a takeoff after an engine ingested sea gulls during takeoff. During deceleration, the affected engine disintegrated and caught fire. Several tires and wheels also disintegrated and the right landing gear collapsed as the aircraft rolled to a stop. Fire erupted on the right wing.

No seats failed but passengers reported that ceiling panels were dislodged, oxygen mask access doors opened, a movie projector came down, and carry-on baggage, pillows, and blankets were thrown into the aisles. Even though all this hampered the evacuation, the burning aircraft was emptied within one minute with only two passengers receiving serious injuries and 30 minor injuries—many as a result of problems with the escape slides and ropes.

According to the NTSB report, the passengers were surprised at the speed at which fire consumed the aircraft and the short time available for evacuation. Crew and passengers alike indicated that evacuating a full load of 380 passengers would have been impossible.

### Inside the House

A Fairchild FH-227 hit a house about four miles from the airport and came to rest with the nose protruding about 25 feet from the back of the house with most of the fuselage in the basement. Cabin side panels came loose, both overhead racks separated, the galley separated, and most seats separated from the structure at impact. Of the 48 on board, 16 died and 32 were seriously injured. Only one of the survivors escaped from the wreckage without aid, so had there been fire there would have been few if any survivors. The NTSB deemed this one a "severe accident with forces approaching but not exceeding the limits for human tolerance to acceleration for occupants restrained only by a seatbelt."

## No Fire

From those cases, it seems clear that airplanes can be pretty thoroughly wrecked and leave survivors, but fire is a key. So, probably, is the number of passengers on board. If the airplane is full, the chances might be small of getting out after a relatively smooth stop but with the airplane on fire. Other things would affect this, especially the behavior of the passengers. In the DC-10 with airline employees, the evacuation took less than a minute because the passengers had some training. The evacuation of an aircraft loaded with loaded tourists, many of whom might try to retrieve carry-on baggage, would be an entirely different matter.

## Depressurization

Some incidents outlined in the NTSB study on passenger safety education illustrate that people apparently do not pay attention to the safety briefings given on airline aircraft before takeoff.

An L-1011 experienced loss of cabin pressurization while descending from Light Level 290. The cabin altitude increased to 20,000 feet. Most of the oxygen masks deployed as they were supposed to, but 20 oxygen compartment doors failed to open. Passengers donned masks but some placed the mask only over their mouth, instead of over the mouth and nose as covered in the preflight briefing. The flight attendants instructed passengers on the use of the masks, but reported that this was difficult to do while using their oxygen masks.

When a DC-10 depressurized and the cabin altitude rose to

18,000 feet, the oxygen masks deployed properly but only two of the 182 passengers on board donned masks and activated the oxygen flow properly.

Another DC-10 suffered disintegration of an engine. Parts of the engine penetrated the fuselage. The cabin decompressed immediately and one passenger was ejected through a broken window. Some oxygen doors did not open automatically. The cabin altitude reached 34,000 feet; the occupants of the aircraft were subjected to cabin altitudes above 30,000 feet for one minute and to altitudes above 25,000 feet for more than two minutes. There was confusion about how to operate the masks.

Another DC-10 lost cabin pressure and the cabin altitude rose to 25,000 feet. Only two of 53 passengers properly donned the mask and activated the oxygen flow.

The oxygen part of the instructions given before flight is important because, while the crew will always initiate an emergency descent and get the airplane down to an altitude that will sustain consciousness, you sure need the oxygen on the way down.

### Drunk Pilots

One rumor that makes the rounds occasionally is that a lot of pilots are careening around the sky under the influence of alcohol. While this is part of the general aviation accident statistics, it is far less a problem than with automobiles, where as many as half the fatal accidents involve legally intoxicated drivers. In general aviation, the NTSB's figure is 10 percent, which tells us that it happens, but it's a relatively small part of the picture. It's even smaller when you consider that flying after drinking is far more likely to result in an accident than is driving after drinking. Flying is a much more complex task, demanding more of both thinking and motor skills. A marked deterioration in the ability to fly can be found with a small amount of alcohol, less than half that which would make you illegal to drive in most states. In a little research project using a sophisticated flight simulator, a couple of us found that just one good slug of booze resulted in a marked deterioration in both flying technique and adherence to procedures. (We went on with the project, up to a level in excess of the drunk-driving laws, and both concluded that flying an airplane at that level of inebriation would be nothing short of deadly.)

There is no number on this, but is is probably reasonable to assume that 20 percent of the miles driven in the U.S. are driven by people with some alcohol on board, and less than five percent are driven by people over the limit—this small percentage causing 50 percent of the fatal auto accidents. I would bet that less than one-fourth of one percent of the airplane hours flown are by people with a positive alcohol level, and a small percentage of those are flown by people over the limit—this total of positive alcohol levels resulting in 10 percent of the fatal general aviation accidents.

In large airline operations, alcohol has not proven to be a factor in spite of what the movies might lead you to believe. However, in an NTSB study covering the years 1975–1981, toxicological tests were positive for alcohol in 6.4 percent of the tests taken from fatally injured commuter (regional) airline pilots and 7.4 percent in air-taxi operation, so it is a factor there.

Pilots are like anyone else. Few walk under ladders, and most have a twinge of apprehension if a black cat crosses their path when en route to the airport. Friday the 13th is not anybody's favorite day to fly and when coincidences, such as the first officer flying in most accidents or flight deck whistling on recordings are considered, there's always a pause for thought. But flying airplanes is a complex business in which there is a logical explanation for everything. The challenge is to uncover that logical explanation and understand the factors that have led others to err.

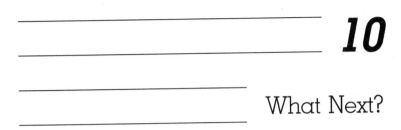

# 10

# What Next?

PREDICTING the future of aviation safety is tenuous at best. Disaster can be predicted and if by chance the prediction comes true, it becomes justified. If there is no disaster, everyone is thankful and nobody really thinks any less of the doomsayer. But there are factors that are working now, factors that *could* have a long-term effect on the general safety of air travel.

## The "System"

The system in which we fly is thought of primarily as the air traffic control system. And when many air traffic controllers went on strike on August 3, 1981, and President Reagan summarily fired them all, many predicted future disaster in the air traffic control system. But the air traffic business worked much more smoothly than anyone imagined. The FAA instituted procedures to meter the flow of traffic, which enabled the relatively small number of controllers left to handle the traffic while new controllers were hired. Some of the procedures that the FAA implemented were an overreaction and some of their facilities basically ignored the dictates coming out of Washington and allowed free access to the IFR system. In other areas, petty bureaucrats reveled

227

in their new-found right to say "you can't do that." But success was in getting through the transition period without a catastrophe, while at the same time providing service that allowed air commerce to continue on a reasonable basis.

After the strike, there were frequent calls to rehire some or all of the fired controllers. The Reagan administration remained adamant on this point, and the FAA was slow in building the controller force back to the prestrike level as almost all users agreed should be done. Traffic still flowed reasonably well, and an FAA program to hold airplanes on the ground for delays rather than stacking them in holding patterns to await approach clearances made the job of the new controllers easier.

One big knot that remains in the air traffic control system needs to be addressed. From a pilot's standpoint, we still see areas where one controller is simply called on to handle too many airplanes. That can only be charged to short staffing. I heard a pilot say, after it had taken him several minutes to get a word in edgewise, "Seems like there are not enough controllers down there." The controller replied, "Oh, maybe its just that there are too many airplanes up there." Some truth either way, but if the system is one that has to restrict use, then it is a failure by any measure.

The controllers who went on strike were good folks. I remember one, smarting over people who said he was un-American for striking against the government, reminding me that he had a son at West Point. They were dedicated, and felt justified in walking out. The management of the FAA had its arrogant moments (and still does) and it wasn't difficult to sympathize with the strikers—especially the ones who walked to demonstrate a point and who would very much have liked to go back to work in a few weeks. All that is apparently water under the bridge, though, and the new lot of controllers is dedicated; it is a disservice to them to say that the system is not as good now as it was before the strike. They don't have as much experience yet, and sometimes when flying you can notice little things that reflect this, but on balance the system is in very good shape. Anyone who predicts a disaster might be proven correct, but there is no more chance of it happening now than in 1980.

## No Crowds

An important thing to understand about the air traffic control system is that the sky itself is not crowded. You can often fly all the way across the country and, while en route, seldom see an airplane. The crunch is at airports. Billions could be spent on radar and computers in the New York area, for example, but until the number of usable runways is increased, the capacity of the system cannot be increased. Often on the commuter train platform I chat with buddies who relate tales of extensive delays getting back into one of the New York airports the night before, when I cruised into my home base of Trenton, New Jersey, in my Cessna without a moment's delay. Their delay was not caused by too many airplanes in the sky, or by my small airplane slipping into Trenton, but by the fact that too many airliners wanted to arrive at one of three airports in New York in too short a span of time—and they can only land one at a time at each airport. The FAA, for safety's sake, meters the flow into those overcrowed airports to maintain the highest possible level of safety. Again, the only way to address this problem is with more runways— preferably at a new airport that is far enough away from existing ones so that the patterns of arriving and departing traffic do not overlap. Until this is done there is no way to increase capacity without seriously compromising safety.

## The Fuzz

Another part of the system is the FAA's inspector force—the people that oversee the certification and operation of airplanes and airlines and the certification of pilots. While many complain about FAA intrusion, the organization often borders on being benevolent. In accident reports, the NTSB often cites some shortcoming of the FAA as being a factor. Perhaps one reason for this is that the FAA has, over the years, grown ever more bureaucratic and the employees have to spend so much time piddling with internal make-work that they have less time to get out into the field and do their proper job of inspecting and observing. For a fact, it often takes an accident to cause the FAA to come down hard on an airline, which suggests that enforcement is more reactive than preventive in nature.

There is strong pressure on the FAA to improve the monitoring of airlines, though, and this should bear some fruit. I always wondered why, when they fine an airline or other user, that money doesn't go into a special fund used to hire more inspectors. That would certainly give the users some incentive to fly the straight and narrow. Get caught, get fined, and that means more inspectors probing around the operation!

The administrator of the FAA is a political appointee, someone to be a figurehead and take the heat. The organization is really run by the second level of management—career people who watch administrators come and go. On balance, these people do a pretty good job of keeping the FAA on track, promoting

both aviation and aviation safety. They are helped by communication from the users of the system, the pilots.

## Deregulation

Deregulation of the airline industry has been hailed as a boon
to the airline traveler, and it certainly has resulted in lower fares
for some seats on some routes. But are cheap seats as safe as more
expensive seats? The cattle-car aspects of travel on some of the
new low-fare carriers are obvious to the passenger, and that's
something that a passenger can voluntarily take or leave. But
when a person buys a ticket on an airline, that person has every
right to expect the same low-risk flight as if he had paid a higher
fare.

Only history will be able to judge this. Many have been
predicting disaster on the new or low-fare carriers. There have
been some accidents and inevitably there will be more. When
there is a spectacular one, the chances are the pilots involved
will be less experienced than those on old-line established carriers,
and the airplane might well be quite old. Those factors have been
true in the "new" airline accidents that have already occurred.
But you have to remember that there were wrecks before deregulation, and those airplanes were crewed by highly paid and
experienced crews flying airplanes that were, on balance, probably younger than those flown by the current crop of discount
carriers.

The primary thing to consider about deregulation is money,
because money has to be at least part of the basis of safety. It's
money that pays and keeps happy the flightcrews and mechanics.
Money is necessary to buy new equipment. And profit is necessary

to keep the airline viable and prompt the management philosophy that puts safety ahead of everything else.

Using a table of operating costs in *Aviation Week and Space Technology* magazine, the numbers for different airlines are found to vary widely. Using the Boeing 727-200 for an example, in the third quarter of 1985 the direct operating cost per hour was found to range from a low of $1,642 per hour (Continental) to as high as $2,317 per hour (TWA). However, when maintenance cost is considered, Continental spent $345.20 per hour compared with $381.10 for TWA. Both airlines used about the same amount of fuel so there wasn't much difference there. That leaves only the crew: Continental, $205.40 per hour versus TWA, $590.20 per hour. That is a *big* difference and it is readily apparent why airline management is leaning heavily on flightcrews in expense-reduction programs.

The question is whether or not the lower paid flightcrews fly as safely as the higher paid flightcrews. There is no doubt that they want to. There is no question that they receive the proper training. One major airline ran an ad once that suggested that their cockpit experience was higher than a cut-rate airline but this was widely viewed as hitting below the belt and the ad disappeared. But this is still in the back of the minds of people. One school of thought holds that flightcrews reached the point where they were grossly overpaid, and the new salary schedules only reflect reality. The other school of thought is that you are in deep trouble when a $50,000-a-year captain gets to a $100,000 thunderstorm. But however you slice it, the experience level on airline flight decks is going to have to decrease as new airlines are formed, as old airlines expand, and as those graybeards that many of us love to fly with retire. The challenge is in the initial and recurrent training processes, as well as in management philosophy, which

many think has more to do with flying safety than any other
factor.


## Per Mile

The more telling, and perhaps bothersome, number is in how
much money it costs to fly one seat one mile. For Continental,
in the period previously mentioned, the cost was 2.5 cents per
seat mile for direct operating cost. The total expense, which
would include depreciation on equipment, insurance, and other
overhead, is at least twice the direct cost. In the case of Continental, a *Newsweek* report tagged their total cost at 6.1 cents a
mile. That would be for the entire fleet; let's say for the sake of
this exercise that it's a nickel for their 727-200s, or double the
direct operating cost. And it would be highly unlikely that anyone
could do it for less than five cents a seat mile.

Fly from New York to Denver for $69? That's 1,640 miles
and the simple math makes it into a 4.2-cents-a-mile fare, or less
than the total cost of operating the airplane even if all the seats
are full. It does cover the direct operating cost, and might allow
for a positive cash flow where only direct costs are considered,
but it clearly puts nothing away for the future, when the airplanes
are all worn out and need major work or when new ones must
be purchased. It also allows nothing for less than a full airplane,
and while the low-cost carriers may seem always packed to the
gills, they can't do this on all flights. In fact, the average load
factors of most of the low-cost carriers is only marginally better
than that of the other airlines.

As airlines merge and jockey for position, they will always

target some routes for low fares, for competitive reasons. But this can't go on forever, and the safety implications of it could some-day come to the surface as the equipment gets older and replacing it with new equipment is financially out of the question. (I would add that Continental, which was used as an example, has been able to put some new equipment on the line even while it was operating in Chapter 11 bankruptcy.) It is possible to ride for not much money today, but airline passengers will eventually pay more, to reflect the real total cost of safely operating the airplanes.

## Hub and Spoke

Another creation of deregulation has been the development of hub and spoke systems. Where airlines used to have major hubs at major cities—Atlanta and Chicago were foremost for years—hubs are now springing up at places such as Charlotte and Raleigh-Durham, North Carolina, Dayton, Ohio, and Nashville and Memphis, Tennessee, for example. It makes sense from an air traffic control standpoint, because the airlines are far less likely to encounter delays at these places than at, say, Atlanta. Less traffic means less exposure to a possible glitch in the system, so this could be positive from a safety standpoint.

You start out, fly to a hub, change planes, and fly out to the final destination. Back in the good old days, there was probably a through flight and perhaps a nonstop. If the latter was the case, then the hub-and-spoke system added a little blip of risk to the trip because most airline accidents happen during takeoff, ap-proach, and landing. Add a landing and add a little risk.

## Aging Equipment

The effect of age on airplanes was mentioned as a factor. In theory, this should not matter so much, given good maintenance, but the fact is, the older an airplane gets the more expensive the maintenance becomes. And when you come to the costly items, such as engines, the old airplane with new engines might not be worth any more money than the engines themselves cost. This is becoming a strong factor in general aviation airplanes, and could come to affect airline airplanes. There are now a lot of business jets, turboprops, and piston-engine twins that are flown until the engines have reached time for overhaul or replacement, and then the airplanes are junked. Or, in some cases, the engines are overhauled at absolutely minimum cost. This is the scary part. The tendency would be to put as little money into an older airplane as is legally required, and in flying operations a lower level of risk is very definitely available through exceeding the minimum legal requirements.

## The Total Key

Aviation is not an old activity. It started in this century in powered airplanes and putted along at a relatively low level until the 1930s, when the DC-3 revolutionized air travel. Since then, all advances have been evolutionary except for Concorde, which was revolutionary as the single largest gain in speed history. Evolution doesn't come without some pain, though, and along the way there have been errors in designing, building, and flying airplanes, just as there have been errors in everything from bridge and building design to errors of judgment in politics.

When Challenger exploded, it was a clear reminder to all who fly that things can and do go wrong and that no process is infallible. The total resources of our land seemed to be riding Challenger, and faith in our technology was shaken by that tragedy, and by the investigation that followed.

But, as in airplane accidents, the tragedy was sudden and final, and lessons must be learned and used to insure that it doesn't happen again. The more we fly, and the more we learn, the more risk we can remove from flying. Those of us who work with aviation safety matters feel reasonably good about things now, but I hope we all realize that the flight that really counts in the *next* one. The reality is that there are things that can happen that have not yet happened. The challenge is to try to anticipate what they might be and head them off at the pass.

# Index

ability of pilot, 11–14
aborting of takeoff, 23
accident rates: and type of airplane, 150; single- and twin-engine airplanes, 161
accident reports, study of, 28–29, 78, 210–11
age of equipment, and safety, 235
airborne weather radar, 18, 98–100
airborne wind shear alert system, 98, 101, 122
Air Force One, 191
Air Force Two, 191–94
airframe failures, 167–69
*Air Line Pilot*, on twin engine overwater flights, 173–74
airline pilots, and approach accidents, 129. *See also* pilots
airlines: accidents, 13, 219; alcohol use, 225; discipline of pilots, 14; minimum descent altitudes, 42; relative safety, 1–3; responsibility of pilots, 58; rules of, 57–58; and see-and-avoid practices, 188
airline transport pilot certificate, 57
airplanes: certification standards, 149–50, 154; failure of, 31

Airport Radar Service Areas, 39; and VFR flight, 180, 181
airport surface detection radar system, 205
airport traffic areas, speed limits, 39
airport vicinity thunderstorms, 47–57
air pumps, maintenance of, 166
air route traffic control center, 182
airspeed indicators, faulty, 69
air taxis, 6; alcohol and, 225
air traffic congestion, 229–30; and collisions, 180; and decision-making process, 120
air traffic control, 34, 35–38, 100–101, 180–81, 211, 227–30; Atlanta airport, 206–209; conflict alert function, 192
air traffic controllers: and bad weather, 49; faulty judgment, 192; strike, 227–29
air traffic radar, and weather, 100
airworthiness directives, 156
alcohol: and pilot error, 73; and safety, 224–26
alternator systems, 166
altimeters, 131
altitude reporting transponders, 200